ibvt-Schriftenreihe

Schriftenreihe des Institutes für Bioverfahrenstechnik
der Technischen Universität Braunschweig

Herausgegeben von Prof. Dr. Rainer Krull

Band 82

Cuvillier-Verlag
Göttingen, Deutschland

Herausgeber
Prof. Dr. Rainer Krull
Institut für Bioverfahrenstechnik
TU Braunschweig
Rebenring 56, 38106 Braunschweig
www.ibvt.de

Bibliographische Informationen der Deutschen Nationalbibliothek
Die Deutsche Nationalbibliothek verzeichnet diese Publikation in der Deutschen Nationalbibliographie; detaillierte bibliographische Daten sind im Internet über *http://dnb.d-nb.de* abrufbar.
1. Aufl. – Göttingen: Cuvillier, 2020

Nonnenstieg 8, 37075 Göttingen
Telefon: 0551-54724-0
Telefax: 0551-54724-21
www.cuvillier.de

1. Auflage, 2020
Gedruckt auf säurefreiem Papier

ISBN 978-3-7369-7211-7
eISBN 978-3-7369-6211-8
ISSN 1431-7230

Quantitative analysis of the electrochemically active bacteria *Geobacter sulfurreducens* and *Shewanella oneidensis*

Von der Fakultät für Maschinenbau

der Technischen Universität Carolo-Wilhelmina zu Braunschweig

zur Erlangung der Würde

einer Doktor-Ingenieurin (Dr.-Ing)

genehmigte Dissertation

von:	Christina Elisabeth Anna Engel
geboren in:	Frankfurt am Main

eingereicht am:	29.04.2019
mündliche Prüfung am:	05.09.2019
Vorsitz:	Prof. Dr.-Ing. habil. Antje Spieß
Gutachter:	Prof. Dr. rer. nat. habil. Rainer Krull
	Prof. Dr. rer. nat. habil. Uwe Schröder

2020

Danksagung

Zunächst möchte ich meinem Doktorvater Professor Rainer Krull danken. Sie haben nicht nur im Studium in zahlreichen Vorlesungen für mein biotechnologisches Vorwissen gesorgt, sondern mir dank der Betreuung des DAAD-geförderten Austauschprogramms mit Waterloo auch die Tür zu einem wundervollen Jahr in Kanada geöffnet. Direkt im Anschluss haben Sie mir die Möglichkeit gegeben innerhalb der Forschungsgruppe ElektroBak am ibvt zu promovieren. Während meiner Promotion haben Sie mir das Vertrauen entgegengebracht unsere Forschung im Bereich bioelektrochemischer Systeme frei und selbstständig zu gestalten, mir aber auch immer mit Rat und Tat zur Seite gestanden, sodass ich in meiner Zeit am ibvt viele neue und wichtige Erfahrungen sammeln konnte. Vielen Dank dafür.

Danken möchte ich auch Professor Uwe Schröder als Kopf unserer Forschungsgruppe ElektroBak. Für Fragen und Diskussionsanregungen rund um Bioelektrochemie hattest Du immer ein offenes Ohr und konntest viele gute Ideen anregen. Danke für Deine Unterstützung beim Paper-Schreiben und Reviewer-Kommentare diskutieren und vor allem vielen Dank für die Übernahme des Korreferats meiner Dissertation.

Der nächste Dank geht an Professor Antje Spieß. Vielen Dank für Deine offene und positive Art unser Institut zu leiten und für Dein Interesse an meiner Arbeit, das oft zu hilfreichen Diskussionen und neuen Anregungen geführt hat. Auch danke ich Dir für die Übernahme des Prüfungsvorsitzes.

An Dr. Katrin Dohnt geht mein besonderer Dank für die Unterstützung und Betreuung während meiner Dissertation. In unseren regelmäßigen Laborbesprechungen konnte ich mit Dir über alle Probleme und Hürden, die zum Promovieren dazugehören, diskutieren und Du hast mir dabei hilfreiche Anregungen für neue Ideen und Lösungsvorschläge gegeben. Auch danke ich Dir für die vielen schönen nicht-fachlichen Stunden beim Plätzchenbacken oder Kochen.

Allen Mitarbeitern des ibvts möchte ich für die angenehme und harmonische Arbeitsathmosphäre danken durch die ich mich am ibvt seit Beginn meiner Bachelorarbeit immer sehr wohl gefühlt habe. Danke an Yvonne und Elena, vor allem für die Unterstützung bei den HPLCs, aber auch bei jeglichen anderen Fragen rund ums Labor. Danken möchte ich Detlev für die Unterstützung beim Bau von funktionierenden MECs. Alex, Dir danke ich für die Einführung in die Welt der Bioelektrochemie und für Deine nette und freundschaftliche Art, die mir den Einstieg ins Doktorandenleben und in mein Thema erleichtert hat. Jan, Dir möchte ich für Deine Unterstützung auf fachlicher Ebene danken und vor allem für Deine Freundschaft und Deine offene Art, die Einbindung in unseren Chor und dass Du immer den richtigen Wein aussuchst.

Ein besonderer Dank geht an meine wechselnden Bürokollegen, die immer ein offenes Ohr für mich hatten und mit denen es nie langweilig wurde. Susanna, o tempo na Spielmannstraße com tu era muito divertido para mim, porque eu podia praticar portugues. Muito obrigada! Antonia, Jeannine, Mathias und Hazel, Euch möchte ich für Eure aufgeweckte und freundliche Art danken, durch die ich mich immer in unserem Viererbüro wohl gefühlt habe. Vielen Dank für Eure Tipps und Anregungen, wenn mal etwas nicht bei meinen Experimenten geklappt hat und für Eure Freundschaft und Unterstützung auch außerhalb vom Labor. Dominik, Lasse und Jonas, Euch möchte ich für Eure muntere Art danken, mit der Ihr durch Euer abwechselndes und manchmal auch gleichzeitiges Auftauchen im „Juniorprofbüro“ dafür gesorgt habt, dass ich in der Endphase meiner Dissertation nicht zum einsamen Einsiedler wurde.

Im Labor haben mich eine Reihe von Studenten begleitet, denen ich für ihre Unterstützung, Ideen und Anregungen und für ihre Arbeiten danken möchte. Theresa, danke Dir für die vielen Stunden im Dunkeln in der Gesellschaft grün leuchtender Biofilme, die Du während Deiner Bachelorarbeit im CLSM-Labor verbracht hast. Marc Stolpmann, Dir danke ich für Deine Ausdauer beim Kampf mit den manchmal widerspenstigen Bioreaktorversuchen während Deiner Studienarbeit. Artem, Dir möchte ich für Deine Unterstützung als HiWi bei meinen Laborversuchen danken. Lena, vielen Dank für Deine nette und aufgeweckte Art und Deine vielen neuen Ideen und Anregungen rund um das Immobilisieren von Bakterien während Deiner Bachelorarbeit. Leonard, vielen Dank für Deine nicht nachlassende Motivation, die elektrochemischen Versuche zum Einfluss des Anodenpotentials während Deines Forschungspraktikums voranzubringen. Jonathan, danke für Deine muntere und aufgeschlossene Art beim Erforschen des Einflusses von Sauerstoff auf unsere Mischkultur während Deiner Bachelorarbeit. Marc Upmann, vielen Dank für Deine aufgeweckte und einfallsreiche Art, die in Lenas Arbeit entstandenen Ansätze und Fragestellungen während Deiner Bachelorarbeit weiterzuentwickeln. Zu guter Letzt geht mein besonderer Dank an David. Danke für Deinen unermüdlichen Einsatz trotz aller Kontaminationshürden während Deiner Bachelorarbeit, die die Grundlagen für unsere Sauerstoffversuche gelegt hat, für Deine Unterstützung bei den nachfolgenden Versuchen als HiWi, für Diskussionen, Anregungen, und Lösungsvorschläge und für den Tipp, das Radio im Labor immer anzuschalten, wenn man mit *Geobacter* hantiert.

Auch in anderen Instituten gibt es viele Menschen, denen ich danken möchte, angefangen mit dem Institut für Mikrobiologie. Simone, Dir möchte ich für die gemeinsame Arbeit an unserem ElektroBak-Teilprojekt danken, für die vielen fachlichen Diskussionen und die gegenseitige Unterstützung beim Entdecken der Bioelektrochemie. Rebekka, vielen Dank für Deine Unterstützung im ElektroBak-Projekt und besonders bei der Anfertigung der Transkriptomanalyse, für das Design unseres Microarrays, die anschließende Auswertung und die Unterstützung beim Paper-Verfassen. Vielen Dank an Annika, für Deine Hilfe im Labor rund um die Transkriptomanalyse, die Einführung in das Arbeiten mit RNA und für die Unterstützung mit dem Microarray. Außerdem danke ich allen BRICS-Mibis, die im 3. OG für eine angenehme Arbeitsatmosphäre gesorgt haben.

Den Mitarbeitern des Instituts für Ökologische und Nachhaltige Chemie möchte ich für die Hilfestellung bei allen Fragen rund um Elektrochemie danken. Danke an Sebastian, André und Igor für Eure Unterstützung bei grundlegenden Fragen zu Beginn meiner Promotion. Vielen Dank an Christopher für die Zusammenarbeit mit Dir im ElektroBak-Projekt und für Diskussionen zu CLSM-Messmethoden. Besonders danken möchte ich Dir, Hilke, für Deine nette und motivierte Art und Deine zahlreichen Anregungen bei unseren regelmäßigen Journal-Club-Treffen. Auch den restlichen Mitgliedern unserer Forschergruppe ElektroBak möchte ich für die nette Arbeitsathmosphäre und die hilfreichen Anregungen während unserer Projekttreffen danken. Besonders danken möchte ich Jana Schlaugat vom Institut für Technische Chemie in Hannover für Deine Unterstützung bei Riboflavinmessungen.

Ein besonderer Dank geht an das Umweltforschungszentrum UFZ in Leipzip an Professor Susann Müller und Florian Schattenberg für die Unterstützung meiner Forschung durch Flow-Cytometrie-Messungen und für die Arbeit und die hilfreichen Anregungen beim Verfassen unserer gemeinsamen Publikation. Vielen Dank an Falk Harnisch und Luis Rosa für hilfreiche Erklärungen rund um Elektrochemie. Ebenfalls danken möchte ich Johannes Dittmann und Professor Markus Böl vom Institut für Festkörpermechanik für die Zusammenarbeit im Bereich mechanischer Beanspruchung von Biofilmen und für den Austausch über CLSM-Messmethoden und Auswertungen.

All meinen Freunden möchte ich für die tolle Zeit danken, die ich mit ihnen verbringen konnte. Nancy und Marlene, vielen Dank für die vielen lustigen Treffen und Wanderausflüge. Danke an alle lachenden Laborläuse für die gemeinsamen Treffen und Roadtrips. Vielen Dank an Esther und Franzi, für unsere regelmäßigen BSRest-Treffen. An Franzi geht ein besonderer Dank für die vielen fachlichen und sprachlichen Tipps beim Korrekturlesen meiner Dissertation. Allen Riesenmikroben danke ich für die vielen lustigen Kochabende! Danke an Lila und Jannik für unsere vielen gemeinsamen Kochabende und die dabei oft ausgelebte Kreativität beim Arrangieren von Liedern. Danke an alle SingDinger für die quirlige Atmosphäre beim Singen, unsere vielen tollen Auftritte und die gemeinsamen Stunden beim Stammtisch und bei Spieleabenden.

Zuletzt möchte ich meiner Familie danken. Besonders danke ich meinem Cousin Volker, für Deine Unterstützung und die vielen, hilfreichen Kommentare beim Korrekturlesen meiner Dissertation. Danke an meinen Bruder Sebastian und meine Schwägerin Estelle für die tolle gemeinsame Zeit mit Euch, egal ob Dresden, Leipzig oder Antwerpen, und an meinen kleinen Neffen Maël, der mit seinem Lachen alle Probleme vergessen macht. Liebe Mama, lieber Papa, Euch danke ich für Eure bedingungslose Liebe und Unterstützung in allen Lebenslagen und dass Ihr mir ermöglicht habt zu dem Menschen zu werden, der ich bin, auch wenn Ihr dafür hinnehmen musstet, dass Euch nie jemand versteht, wenn Ihr versucht zu erklären, was ich eigentlich mache. Ganz zum Schluss möchte ich Benny danken, der mich immer mit freudigem Schwanzwedeln begrüßt, mir manchen Abend beim Schreiben meiner Dissertation auf dem Sofa Gesellschaft geleistet hat und mit seiner frechen und selbstbewussten Art der perfekte Hund auf Erden ist.

Publikationen und wissenschaftliche Arbeiten

Artikel in Fachzeitschriften

Engel C, Schattenberg F, Dohnt K, Schröder U, Müller S, Krull R (2019) Long-Term Behavior of Defined Mixed Cultures of *Geobacter sulfurreducens* and *Shewanella oneidensis* in Bioelectrochemical Systems. Front Bioeng Biotechnol 7:60 . doi: 10.3389/fbioe.2019.00060

Engel CEA, Vorländer D, Biedendieck R, Krull R, Dohnt K (2020) Quantification of microaerobic growth of *Geobacter sulfurreducens*. PLoS ONE 15(1): e0215341 . doi: 10.1371/journal.pone.0215341

Heydorn RL, **Engel C**, Krull R, Dohnt K (2019) Strategien zur gezielten Verbesserung des anodenseitigen Elektronentransfers in mikrobiellen Brennstoffzellen. Chem Ing Tech 91(6):744-757 . doi: 10.1002/cite.201800214

Vorträge

Engel C, Dohnt K, Krull R (2016). Syntrophic interactions during bioelectrochemical degradation of acetate/lactate by a two-species bacterial mixed culture. Plattforminitiative „Mikrobielle Bioelektrotechnologie", 22.-23.11.2016, Braunschweig, Deutschland.

Engel C, Dohnt K, Krull R (2017). Defined mixed cultures of *Geobacter sulfurreducens* and *Shewanella oneidensis* for bioelectrochemical systems. Himmelfahrtstagung „*Models for Developing and Optimising Biotech Production*", 22.-24.05. 2017, Neu-Ulm, Deutschland.

Engel C, Schattenberg F, Dohnt K, Müller S, Krull R (2017). Morphological analysis of electrochemically active biofilms of *Geobacter sulfurreducens* and *Shewanella oneidensis*. "*General Meeting*" der *International Society for Microbial Electrochemistry and Technology* (ISMET), 03.-06.10.2017, Lissabon, Portugal.

Poster

Engel C, Schattenberg F, Dohnt K, Müller S, Krull R (2017). Investigation of the incorporation of *Shewanella oneidensis* in electrochemically active biofilms of *Geobacter sulfurreducens* in a defined mixed culture. Plattforminitiative „Mikrobielle Bioelektrotechnologie", 22.-23.11.2017, Frankfurt am Main, Deutschland.

Studentische Abschlussarbeiten

Vorländer, David (2016). Untersuchungen verschiedener Elektronenakzeptoren mit *Geobacter sulfurreducens* und *Shewanella oneidensis*. Bachelorarbeit.

Briem, Theresa (2016). Optimierung einer Färbemethodik für Konfokale Laser Scanning Mikroskopie zur Morphologieuntersuchung von elektrochemisch aktiven Biofilmen. Bachelorarbeit.

Menßen, Lena (2018). Verbesserung des Elektronentransfers von *Shewanella oneidensis* auf Anoden in bioelektrochemischen Systemen durch Zellfixierung. Bachelorarbeit.

Stolpmann, Marc (2018). Elektrochemische Untersuchung sowie metabolische Charakterisierung von *Geobacter sulfurreducens*. Studienarbeit.

Upmann, Marc (2018). Verbesserung des Elektronentransfers von *Shewanella oneidensis* auf Anoden mittels Fixierung durch Ca-Alginat. Bachelorarbeit.

Block, Jonathan (2018). Sauerstoffeinfluss auf die Stromproduktion elektrochemisch aktiver Mischkultur-Biofilme in mikrobiellen Elektrolysezellen. Bachelorarbeit.

List of Abbreviations and Symbols

List of abbreviations

Abbreviation	Explanation
Ac (index)	acetate
Ag/AgCl (index)	silver/silverchloride electrode
AO	acridine orange
BES	bioelectrochemical system
c-di-GMP	cyclic diguanylate
CA	chronoamperometry
CDW (index)	cell dry weight
CE	Coulombic efficiency
CLSM	confocal laser scanning microscopy
CV	cyclic voltammetry
DO	dissolved oxygen
eDNA	extracellular DNA
EET	extracellular electron transfer
FAD	flavin adenine dinucleotide
FC	fold change
Fd	ferrodoxin
FMN	flavin mononucleotide
Fum (index)	fumarate
HPLC	high performance liquid chromatography
HV	high voltage
Lac (index)	lactate
LB	lysogeny broth
MEC	microbial electrolysis cell
MES	microbial electrosynthesis
MFC	microbial fuel cell
MM	minimal medium
MQ/MQH_2	menaquinon/menaquinol
O2 (index)	oxygen
ox (index)	oxidised
PI	propidium iodide
red (index)	reduced
SHE (index)	standard hydrogen electrode
Suc	succinate
SW	synthetic wastewater
WS	web-source

List of symbols

Symbol	Explanation	Unit
a	area of recorded image	[μm^2]
c	concentration	[$g\ L^{-1}$] or [$mol\ L^{-1}$]
co	surface coverage	[-]
c_{O2}^*	maximum dissolved oxygen concentration	[$mg\ L^{-1}$]
d	biofilm thickness	[μm]
e^-	electron	[-]
E^0	formal redox potential at standard conditions	[V]
$E^{0\prime}$	formal redox potential at pH 7	[V]
F	Faraday constant	[$C\ mol^{-1}$]
k_La	volumetric mass transfer coefficient	[h^{-1}]
L	viability	[-]
n	number	[-]
Q_{real}	transferred charge	[C]
Q_{theo}	theoretically available charge	[C]
$sOUR$	specific oxygen uptake rate	[$mg_{O2}\ g_{CDW}^{-1}\ h^{-1}$]
t	time	[s]
V	volume	[L]
V_B	biofilm volume	[μm^3]
V_d	volume of cells with impaired membrane	[μm^3]
X	biomass concentration	[$g_{CDW}\ L^{-1}$]
$Y_{Fum/Ac}$	yield of fumarate reduced per acetate	[$mol\ mol^{-1}$]
$Y_{O2/Ac}$	yield of oxygen reduced per acetate	[$mol\ mol^{-1}$]
$Y_{Suc/Fum}$	yield of succinate produced per fumarate	[$mol\ mol^{-1}$]
$Y_{X/S}$	yield of biomass produced per acetate	[$g\ g^{-1}$]

Abstract

Electrochemically active bacteria are important biocatalysts in bioelectrochemical systems. This work investigated the long-term behaviour of an electrochemical defined mixed culture of *Geobacter sulfurreducens* and *Shewanella oneidensis*, and the influence of the set anode potential on this defined mixed culture. Further, *G. sulfurreducens* was investigated non-electrochemically in pure culture to elucidate how this bacterium accomplishes oxygen reduction.

Planktonic *S. oneidensis* cells were found to be beneficial for current production and biofilm growth of *G. sulfurreducens* in a microbial electrolysis cell. However, upon removal of planktonic cells, these benefits could not be maintained as *S. oneidensis* was not incorporated sufficiently into the *G. sulfurreducens*-based biofilm. In terms of the applied anode potential, 0.2 $V_{Ag/AgCl}$ as well as -0.2 $V_{Ag/AgCl}$ were found to be best. At 0.2 $V_{Ag/AgCl}$ the highest current density was achieved while at -0.2 $V_{Ag/AgCl}$ the highest current production in relation to biofilm thickness was observed. *G. sulfurreducens* was found to reduce oxygen in non-electrochemical pure cultures at a maximum specific oxygen uptake rate of 95 ± 11 mg_{O2} g_{CDW}^{-1} h^{-1}. The expression of the gene for the cytochrome bd menaquinol oxidase was found to be upregulated under microaerobic conditions, indicating this enzyme to be responsible for oxygen reduction.

Table of Content

1 Introduction

About 10 billion cubic meters of municipal wastewater are being produced every year in Germany (WS-1, 2018). Its treatment requires over 4,000 GWh of electricity, which constitutes roughly 20 % of the municipal energy demand (WS-2, 2018). Aeration of activated sludge tanks has the largest share of energy consumption in wastewater treatment (WS-2, 2018). As worldwide CO_2 emissions are constantly increasing and have reached 33,444,000,000 tons in 2017 (WS-3, 2018), there is both a need to reduce the energy demand as well as to enhance the proportion of CO_2-neutral, renewable energy sources.

One way of combining the need to purify municipal wastewater as well as making a contribution to renewable energy sources and energy savings is to couple wastewater treatment to either the electricity production in microbial fuel cells (MFCs) or the hydrogen production in microbial electrolysis cells (MECs) (Du et al. 2007; Logan et al. 2008). In these bioelectrochemical systems (BESs), a special type of bacteria called electrochemically active bacteria is serving as biocatalyst for the conversion of organic matter to electricity (Logan 2009). As this conversion is an anaerobic process, a large proportion of the energy demand of wastewater treatment plants can be conserved (Gu et al. 2017). At the same time, the energy resulting from the conversion of organic matter is used to produce electricity or the clean fuel hydrogen gas (H_2) (Du et al. 2007; Logan et al. 2008). The way in which electrochemically active bacteria transfer electrons to an anode, which is the ultimate cause of the electricity generation, has been studied for roughly the past two decades and continues to be subject of investigation by researchers around the world.

Understanding how electrochemically active bacteria function on a molecular level and how they interact in complex communities makes it possible to improve MFCs and MECs. Therefore, two very well studied electrochemically active model organisms, *Geobacter sulfurreducens* and *Shewanella oneidensis* (Caccavo et al. 1994; Venkateswaran et al. 1999), were used in this work to investigate a few specific research questions:

- What is the long-term behaviour of interactions between electrochemically active bacteria in a mixed culture?
- Which anode potential provides the optimal conditions for bacterial performance in MECs?
- How does *G. sulfurreducens* handle oxygen intrusion into its living environment?

By trying to answer these questions, this thesis intends to provide some further knowledge for the improvement of BESs.

1.1 Redox reactions – the energetics of electron transfer

Redox reactions occur between two chemical species where electrons are transferred from one species in a reduced state (*red*) to another in an oxidised state (*ox*) (Binnewies et al. 2004). They consist of two half-reactions, the reduction and the oxidation (Binnewies et al. 2004). The removal of a number of *n* electrons (e^-) from one species (*A*) is called the oxidation (equation 1.1) while the uptake of the electrons from the second species (*B*) is called the reduction (equation 1.2) (Binnewies et al. 2004).

$$\text{Oxidation: } A_{red} \rightarrow A_{ox} + ne^- \quad (1.1)$$

$$\text{Reduction: } B_{ox} + ne^- \rightarrow B_{red} \quad (1.2)$$

Electrons present in a reduced species have a certain energy level (Binnewies et al. 2004). This energy level is quantified by the reduction or redox potential $E°$ (Binnewies et al. 2004). A high energy level is represented by a low redox potential and vice versa (Binnewies et al. 2004). The redox potential is not an absolute value, but rather has to be reported in comparison to another half-reaction (Binnewies et al. 2004). The half-reaction $2H^+ + 2e^- \leftrightarrow H_2$ has been defined to have a redox potential of $E° = 0\ V_{SHE}$ (SHE = standard hydrogen electrode) and by default all other potentials are reported in relation to it (Binnewies et al. 2004). Redox potentials are dependent on temperature, pH and the concentration of as well as ratio between reduced and oxidised species (Binnewies et al. 2004). Therefore, standard reaction conditions are defined for this half-reaction which are H_2 at a partial pressure of 1,000 hPa, a proton activity of 1 mol/L and a temperature of 25 °C (Binnewies et al. 2004). In biological systems, it is more common to report redox potentials at neutral pH, which is indicated by an apostrophy ($E°'$) (de Bolster 1997; Acworth 2003).

Transfer of electrons from species A to species B will only occur spontaneously if the energy level of electrons in B_{red} is lower than the energy level in A_{red} (Binnewies et al. 2004). In other words, electrons are only transferred spontaneously from components with a low redox potential to components with a higher redox potential (Binnewies et al. 2004). This principle is shown in **Figure 1.1**. When a redox reaction takes place, the difference in energy levels of electrons between the two reaction partners is converted into other forms of energy, for example heat, new chemical bonds or electricity (Binnewies et al. 2004; Nelson and Cox 2017). Every living organisms is making use of this potential energy from redox reactions as shown schematically in **Figure 1.1** (Acworth 2003). In general, electrons residing in substrates used by cells have a low redox potential (Nelson and Cox 2017). Substrates are converted intracellularly by enzymes, whereby electrons are transferred in multiple reaction steps until they are finally transferred to a terminal electron acceptor with a high redox potential (Nelson and Cox 2017). For example, in aerobic organisms the terminal electron acceptor is oxygen (Nelson and Cox 2017). The half-reaction $O_2 + 4H^+ + 4e^- \rightarrow 2H_2O$ has a redox potential of $E^{\circ\prime} = 0.815\ V_{SHE}$, thus the energy gain from the transfer of electrons to this acceptor is high (Wood 1988). The transfer of electrons is catalysed by a number of enzymes residing in the bacterial membrane which use the potential energy from the electron transfer to translocate protons from the cell interior to the periplasm (Nelson and Cox 2017). The thus created proton gradient is converted into chemical bond energy, usually in the form of ATP (see **Figure 1.1**) (Acworth 2003; Nelson and Cox 2017).

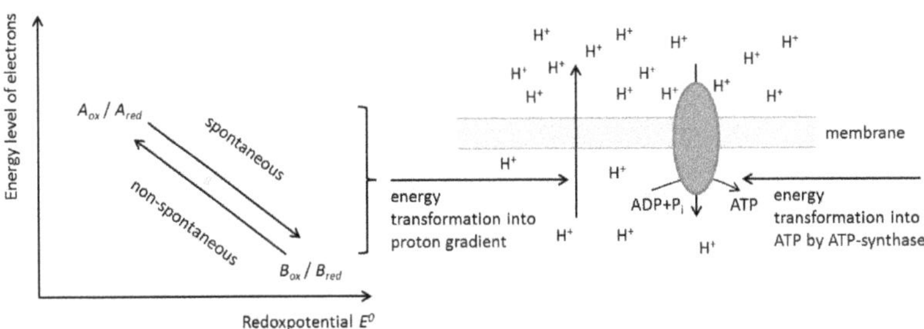

Figure 1.1: Scheme of energy conversion from substrates metabolised via redox reactions into a proton gradient and subsequently into ATP.

1.2 Bioelectrochemical systems – making use of the energy from redox reactions

In bioelectrochemical systems (BESs) the two half-reactions of a redox reaction are occurring separately at an anode (oxidation) and a cathode (reduction). In contrast to classical electrochemical systems, the redox reactions are in part carried out by microorganisms. In general, there are two different types of BES, microbial fuel cells (MFCs) which produce energy, and microbial electrolysis cells (MECs) which require an additional energy source (see **Figure 1.2**) (Lovley 2006; Logan et al. 2008; Logan 2009).

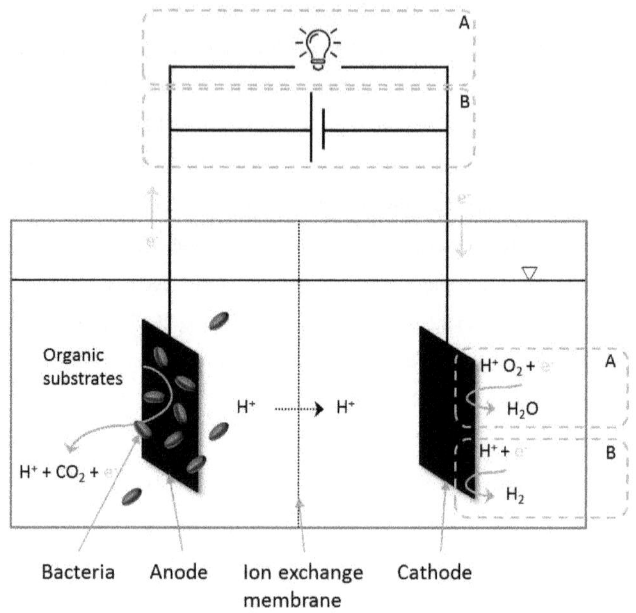

Figure 1.2: Scheme of (A) microbial fuel cell (MFC) and (B) microbial electrolysis cell (MEC) (adapted from Lovley (2006) and Logan et al. (2008)).

In MFCs, bacteria are used to produce electricity from the energy stored in organic substances (Lovley 2006; Logan 2009). Hereby, bacteria reside in the anodic compartment of the MFC and transfer electrons to an anode (see **Figure 1.2 A**, Lovley, 2006; Logan, 2009). Electrons are then conducted through an external electric circuit to the cathode compartment, where the reduction of e.g. oxygen to water is occurring (Lovley 2006; Logan 2009). The potential difference between these half-reactions is positive so that electrical energy is gained (Lovley 2006; Logan 2009). The prospect of MFCs lies in their ability to purify complex wastewater streams while at the same time turning the therein contained chemical bond energy directly into electricity (Bond and Lovley 2003; Logan 2009; Hallenbeck et al. 2014). Anodic and cathodic compartments have to be separated to ensure that bacteria transfer electrons to an electrode

(Lovley 2006; Du et al. 2007). Otherwise, bacteria would transfer electrons directly to oxygen without delivering part of the energy stored in organic substrates in the form of electrical energy (Lovley 2006). A separation is usually realised with a membrane that allows protons to diffuse to the cathodic chamber (Lovley 2006). The membrane constitutes an internal resistance for the system, which can create limitations for MFC performance (Logan 2009; Hallenbeck et al. 2014).

In MECs, electrons released by bacteria are used for other reactions, for instance the production of H_2 from protons (see **Figure 1.2 B**). The redox potential of the reaction $2H^+ + 2e^- \leftrightarrow H_2$ is low (-0.414 V_{SHE} at pH 7) so that the half-reactions taking place at the anode and the cathode usually have a negative potential difference (Logan et al. 2008). This means they are not spontaneous (Logan et al. 2008). Rather, an external voltage has to be applied in these systems that provides the energy for the reaction (Logan et al. 2008; Rosenbaum et al. 2010). The need of additional energy to drive the reaction forward also means that the reaction will not take place spontaneously within the liquid medium even if anode and cathode compartments are not separated (Logan et al. 2008). Thus, the resistant, separating membrane can be eliminated (Logan et al. 2008).

Mostly, H_2 is produced by electrolysis of water. This process requires a high cell voltage of 2.3 V on average to be applied to an electrolysis cell (Call et al. 2009). In MECs operated with acetate or lactate, the theoretical amount of voltage needed is only 0.14 V or -0.011 V, respectively (Rosenbaum et al. 2010). In practice, higher potentials of about 0.5 V to 0.7 V are needed to achieve worthwile H_2 production rates, but this still constitutes a huge energy conservation compared to classical electrolysis of water (Call et al. 2009; Rosenbaum et al. 2010).

It is also possible that, instead of the anode, bacteria are present at the cathode where they take up electrons and use them to form reduced, high-energy compounds, often by CO_2 fixation (Hallenbeck et al. 2014; Rosenbaum and Franks 2014). This is called microbial electrosynthesis (MES) (Rosenbaum and Franks 2014). Usually the anodic half-reaction of MES is the cleavage of water to protons and oxygen (Rosenbaum and Franks 2014). The potential of MECs lies in their ability to store electrical energy in a chemical form. This is advantageous for power plants with inconsistent productivity like solar arrays or wind turbines (Rosenbaum and Franks 2014).

1.3 Electrochemically active bacteria

As stated in chapter 1.1, bacteria use redox reactions to gain energy for cell metabolism and growth by transferring electrons present in the substrates they consume to a terminal electron acceptor (Nelson and Cox 2017). Typically, terminal electron acceptors are soluble molecules that are reduced intracellularly (Babauta et al. 2012a). But there are also bacteria capable of transferring electrons through their cell membranes and periplasm to an insoluble electron acceptor (Babauta et al. 2012a). Bacteria capable of performing this extracellular electron transfer (EET) are termed electrochemically active bacteria (Babauta et al. 2012a).

Natural insoluble electron acceptors encountered by electrochemically active bacteria are different forms of metal oxides often found in sediments of water bodies (Santos et al. 2015). Electrochemically active bacteria use in principle the same transfer mechanisms to reduce a number of different metal ions (Santos et al. 2015). In terms of these transfer mechanisms, anodes in bioelectrochemical systems are quite similar to insoluble metal oxides (Babauta et al. 2012a). Therefore, anodes can serve as terminal electron acceptor to bacteria, which allows the production of electricity.

Among all electrochemically active bacteria, two have been studied in detail due to their ability to respire insoluble metal oxides (Hallenbeck et al. 2014). The first and possibly most important one is *Geobacter sulfurreducens*, which belongs to the δ-proteobacteria (Caccavo et al. 1994). In MFCs inoculated with wastewater that contains an undefined mixture of hundreds of microorganisms, *Geobacter* spp. quickly become dominant within the anodic biofilm, which is the reason why *G. sulfurreducens* has been studied by many researchers (Logan 2009). *G. sulfurreducens* was described first by Caccavo et al. (1994) to be an obligately anaerobic, non-motile, Gram-negative, rod-shaped bacterium. It is capable of reducing different forms of Fe(III), Mn(IV), U(VI), elemental sulfur, fumarate and malate (Caccavo et al. 1994; Mehta et al. 2005; Shelobolina et al. 2007). As carbon and energy source, *G. sulfurreducens* utilises acetate (Caccavo et al. 1994). Other possible electron donors are H_2 and lactate (Brown et al. 2005; Call and Logan 2011). Unlike most bacteria, *G. sulfurreducens* is capable of oxidising acetate completely to CO_2 under anaerobic conditions, which makes this bacterium especially interesting for the use in wastewater-treatment-MFCs as a complete degradation of organic substrates is the aim in wastewater treatment (Bond and Lovley 2003).

Another well-studied organism is *Shewanella putrefaciens* MR-1, which was found in 1988 by Myers and Nealson (1988) and reassigned *Shewanella oneidensis* in 1999 (Venkateswaran et al. 1999). It belongs to the γ-proteobacteria and can reduce iron, manganese and uranium oxides as well as elemental sulfur (Venkateswaran et al. 1999). Possible electron donors are lactate, succinate and fumarate (Venkateswaran et al. 1999). *S. oneidensis* cells are Gram-negative and rod-shaped, but unlike *G. sulfurreducens* this bacterium is a facultative anaerobe (Venkateswaran et al. 1999). Under aerobic conditions, the complete oxidation of the carbon source to CO_2 is possible, but under anaerobic conditions, acetate is the fermentative end product from lactate or pyruvate (Meshulam-Simon et al. 2007). Cell growth under anaerobic conditions is only supported with lactate as carbon source, but *S. oneidensis* has been shown to be metabolically active and produce H_2 gas when pyruvate serves as the sole energy source (Meshulam-Simon et al. 2007).

Often, electrochemically active bacteria form biofilms to maintain close proximity to their insoluble electron acceptors (Babauta et al. 2012a). Both *S. oneidensis* and *G. sulfurreducens* possess this ability, but *G. sulfurreducens* forms thicker and area-wide biofilms (Babauta et al. 2012a). Electrochemically active biofilms need to be conductive as electron transfer from cells in upper biofilm layers has to be possible (Leang et al. 2013). A high biofilm conductivity is also important for the application of electrochemically active biofilms in MFCs or MECs as it reduces charge transfer resistance at the anode-biofilm interface and results in higher current densities and power outputs (Malvankar et al. 2012; Leang et al. 2013).

1.4 Extracellular electron transfer by *G. sulfurreducens* and *S. oneidensis*

The mechanisms for extracellular electron transfer (EET) have been intensively studied in both *G. sulfurreducens* and *S. oneidensis* (Marsili et al. 2008a; Bond et al. 2012; Pirbadian et al. 2014). While both species display similarities in their transfer mechanisms, there are also fundamental differences (Santos et al. 2015). Generally speaking, EET can be divided into two different mechanisms (Borole et al. 2011; Babauta et al. 2012a). The first is based on c-type cytochromes present in the membrane of bacteria and in the extracellular matrix of electrochemically active biofilms (Babauta et al. 2012a). This cytochrome-based transfer is often termed direct electron transfer due to the close proximity usually maintained between the cells and the electron acceptor (Borole et al. 2011). *G. sulfurreducens* is especially known for its direct transfer mechanisms, in which conductive pili play an important role next to c-type cytochromes (Bond et al. 2012). *S. oneidensis* also uses c-type cytochromes, but it additionally employs a second mechanism that is based on soluble mediators (Marsili et al. 2008a; Breuer et al. 2015). This is also termed indirect electron transfer (Borole et al. 2011). In both cases, electron transfer starts within the cytoplasm of the cells during the metabolisation of organic substrates.

The EET mechanisms of *G. sulfurreducens* are schematically depicted in **Figure 1.3**. Acetate is metabolised by *G. sulfurreducens* via the tricarboxylic acid (TCA) cycle and degraded completely to CO_2 (Galushko and Schink 2000). The hereby released electrons are temporarily stored in the form of NADH. NADH is oxidised by the type I NADH dehydrogenase located in the inner membrane, which simultaneously transports protons into the periplasm, thus creating a proton gradient for ATP synthesis (Mahadevan et al. 2006). The NADH dehydrogenase transfers electrons to menaquinon, which is present within the inner membrane (Caccavo et al. 1994; Galushko and Schink 2000; Butler et al. 2006).

Several proteins residing in the inner membrane (ImcH, CbcL) and at the periplasmic side of the inner membrane (MacA) oxidise menaquinol back to menaquinon to transfer electrons onto c-type cytochromes located within the periplasm (Levar et al. 2014; Santos et al. 2015; Zacharoff et al. 2016). While ImcH was shown to be necessary for the reduction of high redox potential electron acceptors ($\geq 0.0\ V_{SHE}$, (Levar et al. 2014)), CbcL is involved in transfer to acceptors with redox potentials $\leq -0.1\ V_{SHE}$ (Zacharoff et al. 2016). From here, electrons are transferred to periplasmic c-type cytochromes of the PpcA-family, of which at least five different types are present in the periplasm (Santos et al. 2015).

Electrons are transferred from PpcA-family cytochromes through the outer membrane by a porin-cytochrome protein complex that incorporates OmaB or its homologue OmaC (octo-heme c-type cytochrome on the periplasmic side), OmbB/C (a porin) and OmcB/C (dodeca-heme c-type cytochrome on the outer membrane side). OmcB was shown to be necessary for the reduction of insoluble Fe(III) in *G. sulfurreducens* (Leang et al. 2003; Liu et al. 2014). But the deletion of OmcB did not impair current production in BES (Holmes et al. 2006; Nevin et al. 2009). Possibly, the homologue OmcC was able to compensate OmcB in these studies.

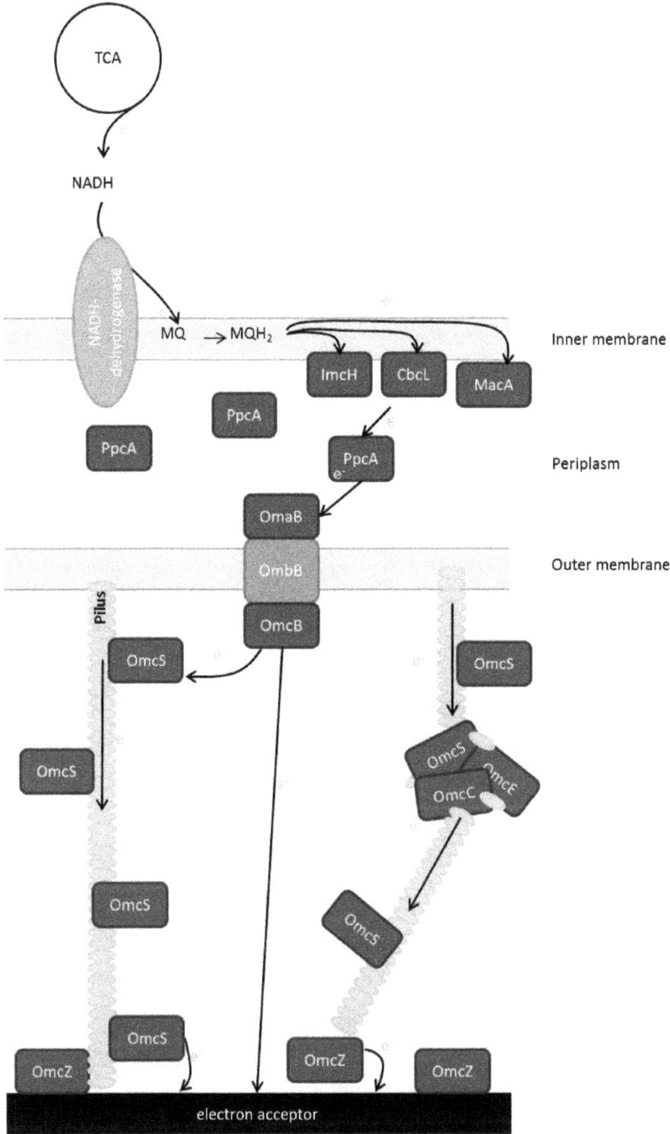

Figure 1.3: Schematic representation of the extracellular electron transfer (EET) mechansims within *G. sulfurreducens* biofilms (Santos et al. 2015; Ordóñez et al. 2016; Reguera 2018).

If a bacterium is closely attached to an anode, electrons can be transferred directly from the outer membrane (Breuer et al. 2015). But since *G. sulfurreducens* forms multi-layer biofilms of up to 200 µm thickness, most cells are not in direct contact with the anode and electron transfer has to be realised through the biofilm (Babauta et al. 2012b). Since electrons can only be transferred from lower to higher redox potentials, there needs to be a redox potential gradient within the biofilm (Babauta et al. 2012b). It has been shown that OmcB is more abundant in cells farther than 10 µm away from the electrode (Stephen et al. 2014). Even when a potential is applied to the electrode which is high enough to allow for complete oxidation of OmcB, a portion of OmcB in these cells will always stay in the reduced state (Stephen et al. 2014). Contrary to this, OmcB located less than 10 µm from the electrode surface is only found oxidised (Stephen et al. 2014). As stated in chapter 1.1, the redox potential is dependent on the ratio between reduced and oxidised forms of a chemical species. Thus, with a greater proportion of reduced OmcB present in upper layers of the biofilm, a redox gradient is maintained ensuring electron flow from upper layers to the electrode (Stephen et al. 2014).

This potential driven electron flow can be realised by the existence of conductive type IV pili composed of PilA in *G. sulfurreducens*. Deletion of the corresponding gene *pilA* led to the inability of *G. sulfurreducens* to reduce insoluble electron acceptors. Thus, it could be concluded that pili are indeed a necessary part in EET (Reguera et al. 2005). The mode of conductivity induced by PilA is still under intensive debate (Lovley 2017; Reguera 2018), but it is quite likely that electron hopping between residues of aromatic amino acids in PilA are responsible for the observed metallic-like conductivity (Reguera 2018). Another c-type cytochrome, OmcS, was found to be located along the pili (Leang et al. 2010). While the spacing of individual OmcS proteins was too high to allow for an explanation of the conductivity of the pili by cytochrome-hopping, deletion of OmcS lead to the inability of *G. sulfurreducens* to reduce insoluble electron acceptors (Mehta et al. 2005; Leang et al. 2010). Thus, OmcS is thought to be responsible for the transfer of electrons from the pili to the insoluble electron acceptor rather than for providing the pili conductivity (Leang et al. 2010). A *G. sulfurreducens* mutant that produced more PilA and OmcS formed more conductive and more stable biofilms on anodes compared to the wild type (Leang et al. 2013). These biofilms also produced higher current densities in potential controlled experiments and higher power outputs in MFCs (Leang et al. 2013). Within biofilms, complexes of OmcC, OmcS, OmcE and other unidentified cytochromes have been found that likely serve as relay for electrons and that are interconnected by PilA (Ordóñez et al. 2016).

The final step for a complete electron transport is the transfer to the anode. Responsible for this is quite likely another well studied c-type cytochrome in *G. sulfurreducens*, namely OmcZ. This c-type cytochrome was found to be located within the matrix of biofilms and is not associated with the outer cell membrane or the pili (Inoue et al. 2011). OmcZ accumulates to high amounts at the electrode surface, filling up the electrode-biofilm interface (Inoue et al. 2011). Therefore it is believed to play an important role in the final transfer of electrons onto the electron acceptor (Inoue et al. 2011).

The electron transfer mechanisms of *S. oneidensis* are in part very similar to those of *G. sulfurreducens*. Electrons are transferred from the inner membrane bound (mena)quinon pool to either CymA or TorC, from which they are transferred to a number of periplasmic multi-heme cytochromes (Breuer et al. 2015). Some of these are quite specifically required for the reduction of fumarate, nitrate or trimethylamine N-oxide in the periplasm (Breuer et al. 2015). A complex that consists of the deca-heme c-type cytochrome MtrA, the β-barrel MtrB and the deca-heme c-type cytochrome MtrC is responsible for transfer across the outer membrane (Breuer et al. 2015). MtrC is found in association with OmcA, another deca-heme cytochrome. Several paralogues of this complex are found in *S. oneidensis* (Coursolle and Gralnick 2012). This transfer pathway is highly similar to the above described mechanism of *G. sulfurreducens* involving inner membrane cytochromes, PpcA-family cytochromes and the Omc-porin complex in the outer membrane.

From the MtrC-OmcA-complex or its paralogues, electrons can be transferred directly to an electrode (Breuer et al. 2015). *S. oneidensis* does not produce conductive pili, but bacteria are capable of forming membrane extensions that contain the MtrABC-OmcA complex and are hypothesised to be used for longer range electron transfer (Pirbadian et al. 2014). If dried, these extensions are conductive (Gorby et al. 2006), most likely due to electron hopping between the multi-heme cytochromes. However, conductivity of these membrane extensions has not been proven under biologically active conditions (Lovley and Malvankar 2015).

In contrast to *G. sulfurreducens*, *S. oneidensis* only forms very thin biofilms and also grows planktonically in BESs (Dolch et al. 2014). For electron transfer across distances too far for membrane extensions, *S. oneidensis* produces soluble electron shuttles. These are different forms of flavins, namely riboflavin, flavin mononucleotide (FMN) or flavin adenine dinucleotide (FAD) (Marsili et al. 2008a; Kotloski and Gralnick 2013). Flavins are reduced by the MtrC-OmcA complex and travel to the anode by diffusion, where they are reoxidised

(Breuer et al. 2015). *S. oneidensis* performs about 75 % of its electron transfer in this indirect manner (Marsili et al. 2008a; Kotloski and Gralnick 2013).

FMN has also been shown to be involved in direct electron transfer by *S. oneidensis*. It forms a complex with MtrC which is referred to as semiquinon and can perform one-electron transfer at 1000-fold higher transfer rates than the two-electron transfer performed by free flavin (Okamoto et al. 2013). An enhancement of current due to riboflavin can also be observed in *G. sulfurreducens*, indicating the possibility that similar complexes also form in this organism, but current increase was only about 10 % compared to cultures without riboflavin supplementation (Okamoto et al. 2014).

1.5 Techniques used to study electrochemically active bacteria

If bacteria transfer electrons to an anode, the anode serves as terminal electron acceptor necessary for cells to gain energy for growth and metabolism (Aelterman et al. 2008; Logan 2009). Therefore, the redox potential of the anode needs to be sufficiently high as electrons can only travel spontaneously to more positive potentials (Logan 2009). In an MFC or MEC, the anode potential varies depending on the type of bacteria growing on the anode, the substrate they use and the redox active proteins they produce to perform EET (Logan 2009). Conversely, the anode potential can influence the metabolic reaction of the bacteria (Aelterman et al. 2008; Bosch et al. 2014). Thus, to study the electron transfer mechanisms, controlled reaction conditions with a defined electrode potential are necessary (Babauta et al. 2012a; Bosch et al. 2014). Electrode potential control is possible by means of a potentiostat (Schroeder and Shain 1969; Logan 2009). Using such a device allows to investigate the influence of specific anode potentials on bacteria and also to detect redox potentials of components necessary for bacterial electron transfer (Aelterman et al. 2008; Bosch et al. 2014).

For electrode potential control, a reference is needed that allows measuring the electrode potential. This is realised with a third electrode introduced into the BES, called the reference electrode. By default, potentials are reported against the standard hydrogen electrode (SHE), but since the defined reaction conditions of this redox reaction are hard to maintain, other electrodes are used as reference in which constant conditions can be maintained in an easier fashion. One very commonly used electrode is the silver/silver chloride electrode (Ag/AgCl), in which the half-reaction $Ag^+ + e^- \leftrightarrow Ag$ takes place. The electrode contains a silver wire that is coated with the highly insoluble AgCl salt. This system is surrounded by a solution containing chloride ions, usually 3 M KCl or saturated KCl. The redox potential *E*°' of Ag/Ag^+ in saturated KCl is at about 0.2 V_{SHE} (Rosa et al. 2017). Potentials measured with this system can be converted into the standard redox potential *E*°' by subtracting this value.

One technique to study the electron transfer capabilities of bacteria in BESs is potentiostatic chronoamperometry. The reference electrode enables a potentiostat to set the potential of one electrode – the working electrode – to a specific and usually constant value. In the case of MFC or MEC studies, the working electrode will usually assume the role of an anode and serve as electron acceptor for bacteria. The other electrode – called the counter electrode and in this case the cathode – serves as electron sink for the reactions taking place at the anodic working electrode. Its potential is controlled by the potentiostat in a way that it will ensure a sufficient consumption of electrons. In this way, current will be limited only by the reactions taking place at the working electrode. This allows for the study of electron transfer by bacteria as all observations made can be referred back to the electrochemically active bacteria present (Babauta et al. 2012a). From the resulting current-time plots (chronoamperograms), the maximum current density achievable by bacteria as well as the amount of charge transferred to the anode can be calculated.

Another technique is cyclic voltammetry (CV), which can be used to determine the formal redox potential of the redox active proteins responsible for the electron transfer from bacteria to an electrode. The working electrode potential is swept at a constant scan rate within a specific potential window in a certain scan direction (from positive to negative potentials or vice versa). If the scan direction goes from positive to negative potentials, this is called a reductive scan as the steadily lower potential of the working electrode will eventually lead to the reduction of redox active proteins present in the surrounding medium. The reduction is observable by a negative current flow at the working electrode. The opposite scan direction will result in a positive current flow at the working electrode caused by the opposite redox reaction and is thus termed oxidative scan. Under substrate depletion, also referred to as non-turnover conditions, reductive and oxidative scans result in current peaks, from which the formal redox potentials of the proteins bacteria use for electron transfer can be determined (Fricke et al. 2008). If substrate is available for consumption by the bacteria, they will create a steady current at high working electrode potentials (Fricke et al. 2008). At more negative values, the potential will be too low for bacteria to efficiently transfer electrons, so the current will decline, eventually reaching 0 A (Fricke et al. 2008). Thus, a sigmoidal curve is created with a sharp increase in current (Fricke et al. 2008). The slope of the current increase is highest when the working electrode potential is that of the redox active proteins responsible for electron transfer (Fricke et al. 2008). Therefore, the first derivative of the current-potential plot will display a peak at the point of the proteins formal redox potential (Fricke et al. 2008).

1.6 Specific research questions regarding the goals of this thesis

1.6.1 Mixed or pure culture – who's the winner?

Research regarding MFCs or MECs is often accomplished with biofilms derived from unspecific mixed cultures, for example wastewater inocula (Logan 2009). In these systems, *Geobacter* spp. are becoming prevalent on anodes due to their superior EET abilities (Logan 2009). Compared to *G. sulfurreducens* pure cultures, unspecific mixed cultures have mostly been reported to have enhanced power outputs (Logan 2009). This indicates positive interactions between *G. sulfurreducens* and other species that result in higher electron transfer rates (Logan 2009).

When comparing *G. sulfurreducens* pure cultures to undefined mixed cultures in an air-cathode MFC, Ishii et al. (2008) observed higher power densities with a mixed culture. Contrary to this, in a setup that used ferricyanide instead of oxygen on the cathode, pure cultures of *G. sulfurreducens* performed better than mixed cultures (Nevin et al. 2008). Thus, one possible explanation for better performance of mixed cultures is the removal of oxygen from the anodic chamber by facultatively anaerobic bacteria. Oxygen may diffuse into the anode chamber from the cathode and is inhibiting current production by *G. sulfurreducens* (Li et al. 2012).

In terms of H_2 production with MECs, there have been studies that showed *G. sulfurreducens* pure cultures may perform just as well as mixed cultures. For example, Call et al. (2009) saw the same H_2 production rates in an MEC from a *G. sulfurreducens* pure culture compared to undefined mixed cultures. Presence of methanogens within the consortium were lowering H_2 outputs by producing methane (Call et al. 2009). This shows that depending on the application, pure cultures are not always disadvantageous.

The positive influences of species in mixed cultures are hard to determine within unspecific mixed cultures as observed effects cannot simply be related to individual species (Logan 2009). Therefore, electrochemically active bacteria have been studied in defined mixed cultures containing only two or three species to reduce the complexity of undefined wastewater inocula. For example, a defined mixed culture of *S. oneidensis* and *G. sulfurreducens* was compared to respective pure cultures by Dolch et al. (2014). The combined cultivation led to a higher number of cells present in the form of an anodic biofilm (Dolch et al. 2014). In another study, *S. oneidensis*, *G. sulfurreducens* and *Geobacter metallireducens* were cultivated in potential-controlled experiments as individual pure cultures and as defined mixed culture (Prokhorova et al. 2017). In the biofilms that formed on anodes, *G. sulfurreducens* was the predominant species, but *S. oneidensis* and *G. metallireducens* were incorporated as well (Prokhorova et al. 2017). The combined cultivation led to an increase in current generation of about 30 % compared to *G. sulfurreducens* in pure culture (Prokhorova et al. 2017). This was explained by an enhanced central carbon metabolism and positive interactions where one species consumes the metabolic end products of another (Prokhorova et al. 2017). Proteins involved in EET were also expressed at higher levels under mixed culture conditions, which explains the higher current production seen (Prokhorova et al. 2017). The studies performed by Dolch et al. (2014) and Prokhorova et al. (2017) only investigated the positive interactions within mixed cultures for timeframes of 7 days or less. Therefore, one goal of this thesis is to investigate the long-term behaviour of a defined mixed culture of *G. sulfurreducens* and *S. oneidensis* with regards to the influence of *S. oneidensis* on the current production and biofilm formation of *G. sulfurreducens* in an MEC.

1.6.2 What is the optimal anode potential?

The anode serves as terminal electron acceptor for electrochemically active bacteria. Therefore, its potential needs to be sufficiently high to allow for the electron transfer from c-type cytochromes or flavins to the anode. The anode potential not only needs to be higher than the redox potential of substrates consumed by bacteria, but also higher than the potentials of the c-type cytochromes or flavins that mediate the transfer. Degradation of acetate to CO_2, for example, has a redox potential of -0.48 $V_{Ag/AgCl}$ (Finkelstein et al. 2006). The multi-heme c-type cytochromes involved in electron transfer display a wide potential window in which they are redox active, ranging from -0.619 $V_{Ag/AgCl}$ to 0.103 $V_{Ag/AgCl}$ (Breuer et al. 2015; Santos et al. 2015). This is due to the multiple heme sides located within these proteins. The wide potential range allows the reduction of a broad spectrum of electron acceptors and thus also on anodes with different potentials (Logan 2009).

The anode potential has been shown to influence the metabolic reaction of the bacteria (Aelterman et al. 2008; Bosch et al. 2014). In principle, a high redox potential would allow bacteria to gain more energy for their cell growth. However, it has been observed by some that the additional available energy is not efficiently turned into biomass (Bosch et al. 2014). Still, a high redox potential should facilitate the transfer of electrons as the potential difference between bacteria and anode, which is the fundamental force for electron transfer, is higher. Ultimately, this could lead to higher current densities due to faster electron transfer rates and higher biomass production rates in BESs. On the other hand, a high redox potential on the anode reduces the efficiency of MFCs as the potential difference between anode and cathode will become lower and therefore power outputs will decline. Similarly, in MECs, a higher external voltage is needed to achieve this high anode potential, i. e. more energy is needed to produce H_2. Thus, the optimal operating potential needs to be determined that balances the positive effects of higher potentials for the bacteria and the negative effects on power or H_2 output. Because of this, the influence of the anode potential has been studied by multiple groups.

Results obtained are, however, quite different (Wagner et al. 2010). Some studies have observed best performance at anode potentials of 0.6 $V_{Ag/AgCl}$ and poorer current production at potentials below 0.1 $V_{Ag/AgCl}$ (Finkelstein et al. 2006; Busalmen et al. 2008). Others have seen the exact opposite, with highest current densities at applied potentials of -0.42 to -0.2 $V_{Ag/AgCl}$ and low current densities at 0.1 $V_{Ag/AgCl}$ or above (Torres et al. 2009; Wei et al. 2010). Results from a number of other studies rather suggest there is an optimum anode potential which is in the range of -0.2 to 0.2 V_{AgAgCl} with decreasing currents from 0.3 $V_{Ag/AgCl}$ onwards (Kumar et al. 2013; Teravest and Angenent 2014; Zhu et al. 2014). The findings of these studies are qualitatively summarised in **Table 1.1**. These different results are seen regardless of whether experiments were performed with pure cultures of *G. sulfurreducens* or *S. oneidensis* or with undefined mixed cultures.

Possible reasons for differences lie in the diverse setups used for experiments. For example, Kumar et al. (2013) found varying best anode potentials depending on whether or not they included a membrane in their system. A clear correlation between the presence or absence of a membrane and the optimal anode potential can, however, not be seen from the other studies. Thus, comparison of results is exacerbated by the fact that all researchers are using different systems for their experiments. Consequently, it appears that the optimal anode potential needs to be determined individually for each system used. Therefore, the second goal of this thesis is to determine the optimal anode potential for the defined mixed culture of *G. sulfurreducens* and *S. oneidensis* in the MEC-setup used here.

Table 1.1: Compilation of results obtained from different studies investigating the influence of set anode potentials on performance in BESs. *G. sulf* = *G. sulfurreducens*, *S. onei* = *S. oneidensis*, mixed = mixed culture, w/o = with or without membrane in BES setup. ++, +, - and -- indicates most positive to most negative results. References: A (Busalmen et al. 2008), B (Zhu et al. 2012), C (Finkelstein et al. 2006), D (Grobbler et al. 2018), E (Prokhorova et al. 2017), F (Kumar et al. 2013), G (Wang et al. 2011), H (Zhu et al. 2014), I (Teravest and Angenent 2014), J (Wei et al. 2010), K (Torres et al. 2009).

		Applied anode potential in $V_{Ag/AgCl}$																
Ref.	Organism, membrane	-0,45	-0,42	-0,36	-0,29	-0,26	-0,2	-0,16	-0,06	0	0,04	0,1	0,2	0,24	0,31	0,4	0,6	0,71
A	*G. sulf*, w/											-					++	
B	*G. sulf*, w/o	--			-					++					++		++	
C	mixed, w/o								-			-					++	
D	*S. onei*, w/o						--						+					++
E	mixed, w/			--		-		++			++			++				
F	mixed, w/o				--		-						++					
G	mixed, w/		-				+			++			++			+		
F	mixed, w/				-		++						--					
H	mixed, w/o	--			-					++					+		--	
I	*S. onei*, w/						--			-			++			-	-	
J	*G. sulf*, w/			++			++						-					
K	mixed, w/		++	++		-						--						

1.6.3 Special characteristics of the metabolism of G. sulfurreducens

Next to electrochemical experiments, basic research focussing on the understanding of the metabolism of bacteria can also help to improve bacterial performance. Because of this, *G. sulfurreducens* was also studied non-electrochemically in the present work, which is why emphasis is given here on some special characteristics of the central carbon metabolism of this bacterium.

The complete oxidation of acetate under anaerobic conditions was already stated above as a special ability of *Geobacteraceae* (Bond and Lovley 2003). The oxidation is happening via the TCA cycle (Galushko and Schink 2000) and is depicted in **Figure 1.4**.

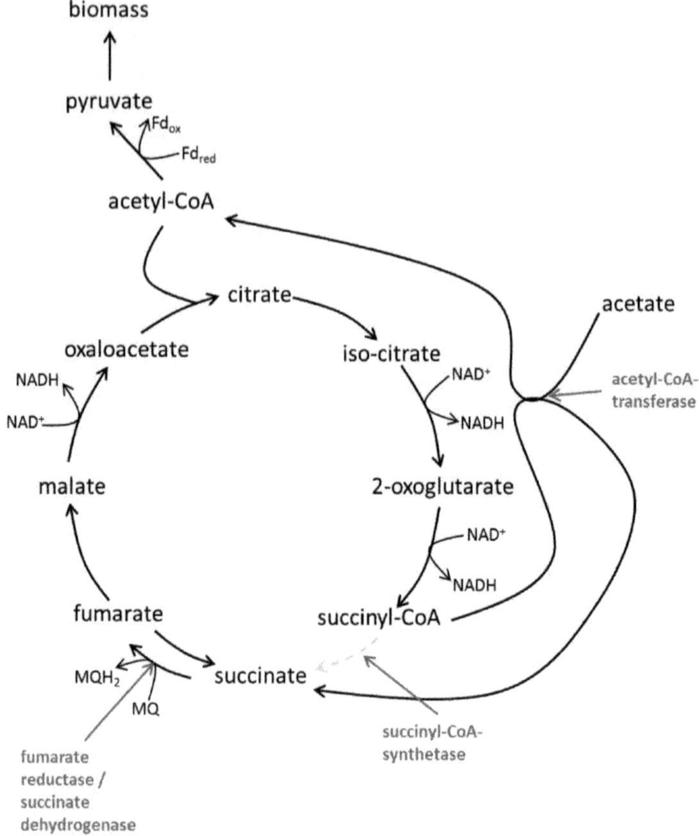

Figure 1.4: Acetate degradation by *G. sulfurreducens* via the tricarboxylic acid cycle derived from Segura et al. (2008) and Butler et al. (2006). Fd = ferrodoxin, MQ = menaquinone, MQH_2 = menaquinol. Red annotations refer to enzymes responsible for the indicated reactions.

Acetate needs to be activated in a reaction catalysed by acetyl-CoA-transferase, in which the CoA-group is transferred from succinyl-CoA to acetate, resulting in succinate and acetyl-CoA (Segura et al. 2008). Acetyl-CoA can now enter the TCA cycle, which functions normally until succinyl-CoA (Segura et al. 2008). There is no succinyl-CoA-synthetase activity in *G. sulfurreducens* as the reaction from succinyl-CoA to succinate is already accomplished by the activation of acetate to acetyl-CoA with acetyl-CoA-transferase (Segura et al. 2008). The next step in the TCA cycle is carried out by a bifunctional fumarate reductase / succinate dehydrogenase (Butler et al. 2006). Depending on the electron acceptor provided, the reaction is carried out in the forward or the backward direction (Butler et al. 2006). If fumarate serves as electron acceptor, it is reduced to succinate and the TCA functions as an open loop whereby succinate accumulates. 75 % of provided fumarate are used as electron acceptor, while 25 % are processed downstream to provide oxaloacetate for acetyl-CoA entry into the TCA cycle (Butler et al. 2006). If soluble Fe(III) or insoluble extracellular electron acceptors are used, the TCA cycle functions as a closed loop (Butler et al. 2006; Mahadevan et al. 2006).

When cultivating *G. sulfurreducens* with either Fe(III)-citrate or fumarate as electron acceptor, an interesting phenomenon can be observed. The reduction of Fe(III)-citrate has a higher redox potential (0.37 V_{SHE}) than the reduction of fumarate (0.03 V_{SHE}) (Butler et al. 2006). Thus, from an energy point of view, cell growth with Fe(III)-citrate should result in higher biomass yields than cell growth with fumarate. However, the opposite is the case. Biomass yields achieved with fumarate are three-fold higher than with Fe(III)-citrate (Esteve-Núñez et al. 2004). The reason for this lies in the mechanism of ATP production used by *G. sulfurreducens*, which is solely due to proton motive force created from NADH dehydrogenase and translocation of protons (Mahadevan et al. 2006). Fumarate respiration takes place intracellularly (Mahadevan et al. 2006). Thus, when electrons are transferred to fumarate within the cytoplasm, two cytoplasmic protons are transferred as well to form succinate (Mahadevan et al. 2006). This increases the difference in H^+ concentration between the periplasm and the cytoplasm which creates the proton motive force for ATP synthesis (Mahadevan et al. 2006). In the case of extracellular Fe(III) as electron acceptor, however, protons remain in the cytoplasm (Mahadevan et al. 2006). This results in an ATP yield that is three-fold lower for Fe(III)-citrate compared to fumarate as electron acceptor (Mahadevan et al. 2006). Similar effects might possibly be observed with other electron acceptors, too.

Finally, another important aspect is the influence of oxygen on *G. sulfurreducens*. In BESs, oxygen intrusion is believed to be one of the reasons why pure cultures of *G. sulfurreducens* do not perform as well as mixed cultures (Logan 2009). Oxygen is thought to hinder the transfer of electrons to an anode by *G. sulfurreducens* (Li et al. 2012). Yet, genome analysis showed the possible existence of enzymes associated with the reduction of oxygen (Methé et al. 2003). These include cytochrome c oxidase, which is complex IV of the oxidative phosphorylation and represents the final step in oxygen reduction carried out by aerobic bacteria (Methé et al. 2003). Two other possible oxygen reducing enzymes whose corresponding genes are found in the genome are a high oxygen affinity cytochrome bd menaquinol oxidase and a rubredoxin-oxygen oxidoreductase (Methé et al. 2003). These findings were surprising as

G. sulfurreducens was originally thought to be a strict anaerobe (Caccavo et al. 1994). Following this, Lin et al. (2004) showed that *G. sulfurreducens* can indeed reduce oxygen. The authors showed that reduction of oxygen even supports cell growth in the same manner as fumarate reduction does. But cells were only capable of consuming oxygen concentrations of up to 10 % in the gaseous head space of cultivation flasks, which is why this ability was not discovered in the original description of the species (Lin et al. 2004).

Since Lin et al. (2004) only observed oxygen concentrations in the headspace, it is not yet known how oxygen concentrations behave within the liquid and at what consumption rates *G. sulfurreducens* reduces oxygen. It is further not yet conclusively understood how *G. sulfurreducens* performs oxygen reduction. Therefore, a third goal of this thesis is to determine the specific rate at which oxygen reduction by *G. sulfurreducens* is possible, to investigate how *G. sulfurreducens* responds to oxygen on a molecular level, and to explore which enzymes are responsible for oxygen reduction.

2 Materials and Methods

2.1 Bacterial strains and cultivation media

For the present work, the wild type strains *Geobacter sulfurreducens* PCA (DSMZ 12127) and *Shewanella oneidensis* MR-1 (ATCC 700550) were used. All work with these bacteria was performed within a biosafety cabinet (Thermo Fisher Scientific, Waltham, Massachusetts, USA) to avoid contamination with other organisms.

The composition of cultivation media is listed below. All chemicals used were at biological or technical grade. All solutions were sterilised at 121 °C and 2 bar for 15 min. Deionised water was sterilised in the same way and used for compounding of media. Heat sensitive components were sterile filtered using cellulose/acetate or regenerated cellulose filters (Sartorius AG, Göttingen, Germany) with a pore size of 0.2 µm.

Lysogeny broth (LB) medium (Bertani 1951) was used for precultures of *S. oneidensis* and was composed as shown in **Table 2.1**. For the preparation of LB-agar plates, agar was added at a concentration of 18 g/L.

Table 2.1: Composition of LB medium.

Component	Concentration [g/L]
NaCl	10
Yeast-extract	5
Peptone	10

For the cultivation of pure cultures of *S. oneidensis*, a minimal medium (MM) was used that was adapted from Hau et al. (2008). The individual solutions used as well as the composition of the final medium are listed in **Table 2.2** to **Table 2.7** below.

Table 2.2: Composition of *S. oneidensis* 10x salt solution

Component	Concentration [g/L]
K_2HPO_4	2.25
KH_2PO_4	2.25
NaCl	4.6
$(NH_4)_2SO_4$	2.25

Table 2.3: Composition of 10 x magnesium sulfate solution

Component	Concentration [g/L]
$MgSO_4$	1.17

Table 2.4: Composition of 10x lactate solution (100 mM)

Component	Concentration [g/L]
Na-L-lactate	11.2

Table 2.5: Composition of trace element solution from Balch et al. (1979) (sterile filtered)

Component	Concentration [g/L]
Nitrilotriacetic acid	1.5
$MgSO_4$ x $7H_2O$	3
$MnSO_4$ x $2H_2O$	0.5
NaCl	1.0
$FeSO_4$ x $7H_2O$	0.1
$CoCl_2$	0.1
$CaCl_2$ x $2H_2O$	0.1
$ZnSO_4$	0.13
$CuSO_4$ x H_2O	0.1
$AlK(SO_4)_2$	0.1
H_3BO_3	0.1
Na_2MoO_4 x $2H_2O$	0.01
Na_2SeO_3 x $5H_2O$	0.0003
$NiCl_2$ x $6H_2O$	0.03

Table 2.6: Composition of vitamin solution from Balch et al. (1979) (sterile filtered)

Component	Concentration [mg/L]
Biotin	2
Folic acid	2
Pyridoxine hydrochlorid	10
Thiamin hydrochlorid	5
Riboflavin	5
Nicotinic acid	5
DL-Calcium pantothenate	5
Vitamin B_{12}	0.1
p-Aminobenzoic acid	5
Lipoic acid	5

Table 2.7: Composition of the minimal medium used for *S. oneidensis*

Solution	Volume fraction [mL/L]
10x *S. oneidensis* salt solution	100
10x Magnesium sulfate solution	100
10x Lactate solution	100
Trace element solution	12.5
Vitamin solution	12.5
Deionised water	675

A minimal medium termed synthetic wastewater (SW) derived from Kim et al. (2005) was used for the cultivation of *G. sulfurreducens* pure cultures as well as mixed cultures of both organisms. All solutions used are listed in **Table 2.4** to **Table 2.11** and the basic composition of the medium is shown in **Table 2.12**. Depending on the cultivation conditions, different amounts of electron donor (acetate or lactate) and electron acceptor (fumarate) were added to this basic composition as indicated in **Table 2.13**.

Table 2.8: Composition of 5x *G. sulfurreducens* salt solution

Component	Concentration [g/L]
NH_4Cl	1.55
KCl	0.65
NaH_2PO_4 x H_2O	13.45
Na_2HPO_4	21.65

Table 2.9: Composition of 10x acetate solution (100 mM)

Component	Concentration [g/L]
Na-acetate	8.2

Table 2.10: Composition of 10x fumarate solution (400 mM)

Component	Concentration [g/L]
Na-fumarate	64

Table 2.11: Composition of 10x cysteine solution (20 mM, sterile filtered)

Component	Concentration [g/L]
L-cysteine	2.42

Table 2.12: Basic composition of synthetic wastewater (SW) medium. x refers to volume added from cysteine, acetate, lactate and/or fumarate solution as of ***Table 2.13***, y refers to volume added from preculture.

Component	Volume fraction [mL/L]
5x *G. sulfurreducens* salt solution	200
Trace mineral solution	12.5
Vitamin solution	12.5
Acetate/Lactate, fumarate, cysteine	x
Liquid *G. sulfurreducens* preculture	y
Deionised water	775 – x – y

Table 2.13: Addition of electron donor acetate or lactate, electron acceptor fumarate, and cysteine depending on cultivation type for *G. sulfurreducens* pure cultures or mixed cultures of *G. sulfurreducens* and *S. oneidensis*.

Cultivation type	Volume fraction [mL/L]		
	Acetate / Lactate	Fumarate	Cysteine
Preculture	100 / 0	100	100 (optional)
Electrochemical cultivations (pure)	100 / 0	0	0
Electrochemical cultivations (mixed)	50 / 50	0	0
Bioreactor cultivations	100 / 0	100 or 10	100

Additionally, a complex SW-medium was used for *G. sulfurreducens* which was composed of the above described basic composition for precultures including cysteine and which was further supplemented with yeast extract at a final concentration of 1 g/L. For SW-agar plates, 18 g/L of agar were added to this complex medium.

2.2 Cultivation conditions for precultures

S. oneidensis cells grown in LB-medium were stored at -80 °C in cryotubes. For precultures, cells from one cryotube were plated on LB-agar plates and incubated overnight at 30 °C. Cells from one colony were used to inoculate 5 mL of LB medium in a sterile 50 mL baffled shake flask (3 baffles) and were subsequently incubated in a shaking incubator (25 mm orbital diameter, Multitron, Infors AG, Bottmingen, Switzerland) at 180 min^{-1} and 30 °C overnight. Alternative to primary plating on LB-agar plates, 100 µL of a thawed cryogenic culture were added directly to 5 mL of LB medium in a baffled shake flask. Cells from a liquid LB-preculture were used to inoculate MM-precultures. 50 mL of MM according to **Table 2.7** were prepared in a sterile 500 mL baffled shake flask (4 baffles). The optical density at 600 nm (OD_{600}) of an overnight grown LB-preculture was determined using a Libra S11 photometer (Biocrom GmbH, Berlin, Germany). From the OD_{600}, the amount of cell suspension was determined that would result in an OD_{600} of 0.1 for MM-precultures. The corresponding volume was transferred into sterile Eppendorf tubes and centrifuged at 13,000 g for 3 min. After removal of the supernatant, the cell pellet was resuspended in MM and transferred into the baffled shake flask. Cells were incubated at 30 °C and 180 min^{-1} overnight (at least 12 h).

G. sulfurreducens cells grown in SW-minimal medium were stored at -80 °C in cryotubes. For cell culturing from the cryogenic state, a cryogenic culture was thawed completely and aliquots of 100 µL were plated on 10 SW-agar plates. SW-agar plates were transferred into an anaerobic jar. Oxygen was removed from the jar with Anaerocult® A (Merck KGaA, Darmstadt, Germany) and plates were incubated at 30 °C. After 10 to 14 days, colonies that formed on the SW-agar plates were used to inoculate a liquid complex SW-medium preculture. 10 mL of complex SW-medium were prepared as described above (chapter 2.1) in sterile 25 mL serum bottles. Medium was gassed with N_2 for at least 20 min to create anaerobic conditions. Subsequently, several colonies from an SW-agar plate were transferred into the complex SW-medium and the serum bottle was closed with a butyl rubber stopper which was secured with a crimp cap. Cells were incubated at 30 °C in a shaking incubator (25 mm orbital diameter) at 180 min^{-1}. After three days, cells from a complex SW-medium preculture were used to inoculate an SW-minimal medium preculture. The OD_{600} of the complex SW-medium preculture was determined and the culture volume was calculated that would result in an optical density of 0.1 in 50 mL of SW-minimal medium. 50 mL of SW-minimal medium including cysteine (see **Table 2.12** and **Table 2.13**) were prepared in a sterile 100 mL serum bottle deducting the beforehand determined complex SW-medium preculture volume and gassed with N_2 for 20 min. Subsequently, the determined complex SW-medium preculture volume was added directly and the serum bottle was closed with a butyl rubber stopper and a crimp cap. Cells were incubated at 30 °C and 180 min^{-1} overnight (at least 18 h).

SW-minimal medium precultures of *G. sulfurreducens* were passaged regularly (approx. every 3 days) by transferring 10 % or 20 % (v/v) (depending on the optical density) of a consisting SW-minimal medium culture into fresh medium as described above. Cysteine was usually omitted from the medium except in the case of precultures that were used for bioreactor experiments (see chapter 2.7).

Regularly, a small volume of a consisting preculture of *G. sulfurreducens* was plated on LB-agar plates using an inoculation loop and incubated aerobically at 30 °C for three days. Hereby, precultures were checked unspecifically for aerobic contamination.

2.3 Setup of microbial electrolysis cell

S. oneidensis and *G. sulfurreducens* were cultivated bioelectrochemically as individual pure cultures and further as a defined mixed culture. Bioelectrochemical experiments were performed in a microbial electrolysis cell (MEC) which is schematically shown in **Figure 2.1**. The MEC consists of a 100 mL flask (Schott AG, Mainz, Germany) equipped with a magnetic stirring bar (2.5 cm length) and closed with a butyl rubber stopper. The stopper was perforated (5 mm diameter) to permit the inclusion of an Ag/AgCl saturated KCl reference electrode (SE11, Meinsberger, Waldheim, Germany). Two graphite plates contacted with silver-coated copper wires (0.6 mm diameter) served as the anode and cathode. Wires were isolated with shrink-on tubes to prevent direct contact with the liquid in MECs. While anodes were 1 cm x 2.5 cm x 0.5 cm, cathodes had a size of 2 cm x 2.5 cm x 0.5 cm. Electrodes were ground with 240-grit sandpaper before use and cleaned with deionised water. Electrodes were inserted into the butyl rubber stopper using cannulae (Sterican® 1.2 mm x 40 mm, Braun, Melsungen, Germany). Assembled MECs were sterilised at 121 °C and 2 bar for 15 min.

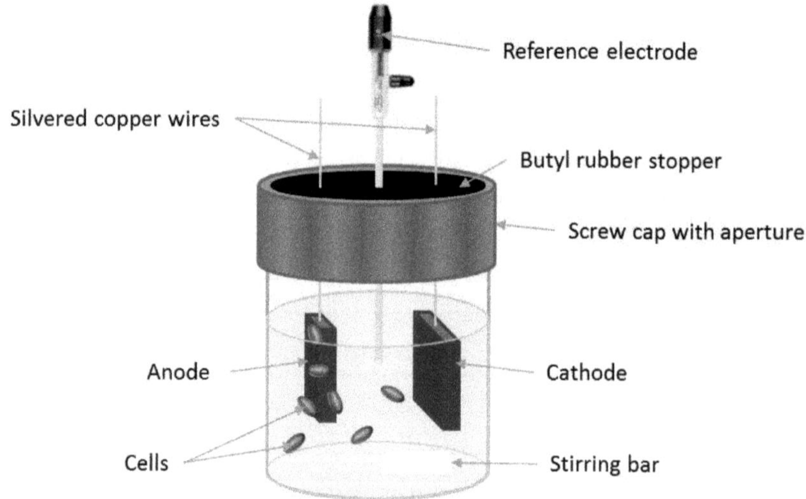

Figure 2.1: Scheme of microbial electrolysis cell used for bioelectrochemical culivations. Source for image of reference electrode: https://www.meinsberger-elektroden.de.

After sterilisation, 80 mL of medium were prepared in MECs depending on the organism used as described above (see **Table 2.7**, **Table 2.12** and **Table 2.13**) and gassed with N_2 for 20 min.

Before insertion, reference electrodes were checked for accuracy by connecting the reference electrode and a second reference electrode of identical type, set aside for this very purpose, to a pH-meter (Knick, Berlin, Germany) and measuring the voltage between the two electrodes in 3 M KCl solution. Reference electrodes were only used if the modulus of the voltage did not surpass 1 mV. Reference electrodes were sterilised using Pursept® AF (Schülke & Mayr GmbH, Norderstedt, Germany) and inserted into the perforation of the butyl rubber stopper. Pursept® also served as lubricant for insertion. At this step of the assembly, the experiment executioner is strongly advised to wear cut-resistant gloves.

After gassing with N_2, MECs were inoculated with cells from minimal medium precultures. The inoculation volume was chosen so that an OD_{600} of 0.1 was achieved in the MEC. In the case of defined mixed cultures, the volume was chosen so that each individual species would result in an OD_{600} of 0.1.

2.4 Electrochemical cultivation and techniques

Prepared and inoculated MECs were connected to a potentiostat (MPG2, Bio-Logic Science Instruments, Seyssinet-Pariset, France). Electrochemical cultivations consisted of three cultivation cycles. In each cycle, two electrochemical techniques, namely chronoamperometry (CA) and cyclic voltammetry (CV), were used. CA served for applying a potential to the working electrode – here the anode – to promote cell growth of bacteria using the anode as terminal electron acceptor. In pure culture experiments, the potential was set to 0.2 $V_{Ag/AgCl}$. The defined mixed culture was carried out at -0.2, 0.0, 0.2, 0.4 and 0.6 $V_{Ag/AgCl}$. Upon usage of the anode as terminal electron acceptor, a current could be measured at the anode, which was recorded by the potentiostat. At the point of highest current (turnover conditions), CV was performed. The working electrode potential was swept at a scan rate of 0.25 mV/s, starting from the potential applied during CA to a potential of -0.45 $V_{Ag/AgCl}$ and subsequently to 0.2 $V_{Ag/AgCl}$ in the case of *G. sulfurreducens* pure cultures and 0.6 $V_{Ag/AgCl}$ in the case of defined mixed cultures. CV scans were repeated three times and terminated at the potential applied during CA. Immediately following CV measurements, CA was continued at the same potential until current subsided. 5x *G. sulfurreducens* salt solution (see **Table 2.8**) was diluted with deionised water to 1x concentration in sterile 100 mL flasks equipped with a magnetic stirring bar and gassed with N_2 for 20 min. Butyl rubber stoppers from MECs containing anode, cathode and reference electrode were transferred into the salt solution and the flasks were sealed. CA was continued for about 12 h, until current dropped to values below 0.002 mA/cm^2. Following this, a second CV was performed (non-turnover conditions) in the same way as described above. This concluded one cultivation cycle. New medium was prepared in sterile 100 mL Schott flasks equipped with a magnetic stirring bar as described above, and butyl rubber stoppers containing anode, cathode and reference electrode were transferred into the fresh medium after non-turnover CV measurement for a second and third cultivation cycle. This procedure was carried out for all cultivations except for the pure culture of *S. oneidensis*, which only comprised one cultivation cycle omitting CVs. During the CA of cycle 3, experiments were terminated and biofilms were used for flow cytometry (see chapter 2.5) and confocal laser scanning microscopy (CLSM, see chapter 2.6) analysis.

During defined mixed cultures and *S. oneidensis* pure cultures, CA was paused occasionally to allow for sampling. MECs were disconnected from the potentiostat. Sterile 2 mL syringes (Braun, Melsungen, Germany) and sterile cannulae (Sterican® 0.8 mm x 40 mm, Braun, Melsungen, Germany) were used to take samples of 1 mL through the butyl rubber stopper without opening the MEC. Immediately following, CA was continued. The OD_{600} of samples was determined. Subsequently, samples were centrifuged at 13,000 g for 3 min. The supernatant was filtered into Eppendorf tubes using cellulose-acetate or regenerated cellulose filters with a pore diameter of 0.2 µm (Minisart®, Sartorius AG, Göttingen, Germany). Samples were stored at -20 °C until HPLC analysis (see chapter 2.10). During pure cultures of *G. sulfurreducens* no such sampling was performed to avoid disturbance of the cultivation by intrusion of oxygen. Rather, samples were only taken at the beginning and end of each cycle before closing and after opening of MECs.

Contamination tests were performed similarly as described above for *G. sulfurreducens* precultures on LB-agar plates. In the case of *G. sulfurreducens* pure cultures, experiments were considered uncontaminated if no colonies formed on LB-agar plates within 3 days. In the case of defined mixed cultures, colonies forming were inspected visually whether or not they were *S. oneidensis*. Other bacteria could be distinguished from *S. oneidensis* by their lighter colour as *S. oneidensis* forms slightly orange colonies.

Data from CA and CV measurements were processed with the software OriginPro (OriginLab, Northampton, Massachusetts, USA). From CA data, the maximum current density was determined. Further, the total amount of charge transferred during CA Q_{real} was determined from the integral of current-time curves for calculation of the Coulombic efficiency (CE). CE was calculated as the ratio of Q_{real} and the theoretically available charge Q_{theo} determined from the amount of degraded substrate Δc. In the case of acetate, the amount of electrons available from the degradation of one molecule is 8 according to the complete oxidation to CO_2 by equation 2.1.

$$\mathrm{CH_3COOH} + 2\,\mathrm{H_2O} \rightarrow 2\,\mathrm{CO_2} + 8\,\mathrm{H^+} + 8\,e^- \qquad (2.1)$$

In the case of lactate, the amount of electrons available from the degradation of one molecule is 12 according to the complete oxidation to CO_2 by equation 2.2.

$$\mathrm{CH_3CHOHCOOH} + 3\,\mathrm{H_2O} \rightarrow 3\,\mathrm{CO_2} + 12\,\mathrm{H^+} + 12\,e^- \qquad (2.2)$$

Thus, the CE can be calculated from this data by equation 2.3, where Δc_{Ac} and Δc_{Lac} are the amounts of degraded substrate acetate and/or lactate respectively, V is the liquid volume of the MEC and F is the Faraday constant (96,485.3365 C/mol) (Binnewies et al. 2004).

$$\mathrm{CE} = \frac{Q_{real}}{Q_{theo}} = \frac{Q_{real}}{(\Delta c_{Ac} \cdot V \cdot 8 + \Delta c_{Lac} \cdot V \cdot 12) \cdot F} \qquad (2.3)$$

CVs were used to determine the formal redox potentials of redox active components (c-type cytochromes or flavins) within the electrochemically active biofilms. In the case of non-turnover conditions, electrons will not be transferred from the working electrode to the component as long as the potential of the working electrode is higher than that of the component, and thus all molecules of the component will be oxidised. Once the working electrode potential is scanned to more negative values reaching the redox potential of the component, electrons will start to be transferred to the component, resulting in current flow. The modulus of current will increase until most molecules in the surrounding of the working electrode are reduced and the further reduction of the component becomes diffusion limited. Thus, a peak in current is created that is located slightly below the formal redox potential of the component. During the oxidative scan, the same principle process is occurring in reverse with a current peak forming that is located slightly above the formal redox potential of the component. From the current peak potentials of the oxidative and the reductive scan, the formal redox potential of the redox active component can be determined by calculating the mean of the two current peak potentials (Fricke et al. 2008). Due to overlying currents, a baseline subtraction was performed to better perceive the peaks.

In the case of turnover CVs, formal redox potentials are determined from peaks obtained from the first derivative of the current-potential curves (Fricke et al. 2008). As several peaks usually overlap, fitting with Gaussian curves was performed to determine between the individual peaks.

An example of the resulting fit can be seen in **Figure 2.2**. The black line represents the first derivative of the current-potential curve with a broad peak located at -0.34 $V_{Ag/AgCl}$. The peak was fitted with three Gaussian curves (blue lines) that result in the cumulative fit represented by the orange line and is in good accordance with the first derivative.

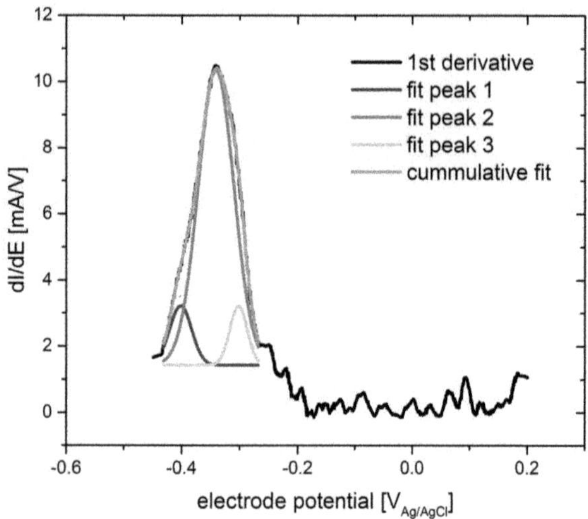

Figure 2.2: Example of peak obtained from first derivative of turnover CVs fitted with multiple Gaussian fits.

2.5 Cell fixation and flow cytometry analysis

After termination of electrochemical cultivations, cells were prepared for flow cytometry analysis as described previously (Zimmermann et al. 2016). For cell fixation, biofilm cells were scraped off the electrodes using a metal inoculation loop and resuspended in 2 mL of PBS solution (8 g/L NaCl, 0.2 g/L KCl, 1.44 g/L Na_2HPO_4, 0.24 g/L KH_2PO_4) in a 5 mL-Eppendorf tube. 5 mL of planktonic cells were likewise transferred into an Eppendorf tube. Cells were centrifuged at 3,200 g and 4 °C for 10 min. The supernatant was discarded and cells were washed twice in 2 mL PBS solution. Cell pellets were resuspended in 2 % paraformaldehyde solution created from an 8 % (w/w) paraformaldehyde stock solution and PBS and incubated at 4 °C for 30 min. Subsequently, paraformaldehyde solution was removed by centrifugation and cell pellets were resuspended in 4 mL of 70 % (v/v) ethanol. The thus fixated cells were stored at -20 °C until further processing.

For flow cytometry measurements, fixated cells were sent to the working group Flow Cytometry (S. Müller, F. Schattenberg), Department of Environmental Microbiology, Centre of Environmental Research (UFZ) in Leipzig, Germany, who kindly performed all further flow cytometry analysis. Prior to analysis, cellular DNA was stained using DAPI as described previously (Koch et al. 2013). DAPI fluorescence and forward scatter signals were measured analysing 250,000 cells per sample residing in the prior determined cell gates for *S. oneidensis* and *G. sulfurreducens*. From this data, the distribution of each cell type within a sample was calculated.

2.6 Confocal laser scanning microscopy

Biofilms were analysed with confocal laser scanning microscopy (CLSM) to determine their thickness and viability. For this purpose, cells were dyed with two fluorescence dyes, acridine orange (AO) and propidium iodide (PI). Both dyes form fluorescent complexes with DNA. While AO can pass through the bacterial membrane of intact cells, PI can only penetrate cells with an impaired membrane (Hobbie et al. 1977; Krämer et al. 2016). A solution containing both dyes at concentrations of 113 µM (AO) and 60 µM (PI) was prepared in optically opaque Eppendorf tubes. Anodes containing biofilms were removed from the butyl rubber stopper of MECs and placed in petri dishes. Biofilms were overlaid with the dye solution and incubated in the dark for 5 min. Subsequently, the dye was carefully allowed to drain from the biofilm. Anodes were overlaid with 1x salt solution prepared by diluting the 5x *G. sulfurreducens* salt solution used for the cultivation medium (see **Table 2.8**).

A Nikon C2+ Confocal Microscope (Nikon, Düsseldorf, Germany) was used for CLSM measurements. Images were recorded using either a 10x air objective or a 16x water-dipping objective. Lasers with a wavelength of 488 nm and 561 nm were used to excite AO and PI, respectively. Laser powers were set to 2.1 % and 0.8 %, respectively. Emission from fluorescence dyes was directed onto the integrated detectors with dicroic mirrors and filters with a bandpass of 525 ± 25 nm and 595 ± 20 nm for AO and PI, respectively. For the AO detector, a High Voltage (HV) of 54 and an Offset of 0 were chosen, while for the PI detector an HV of 86 and an Offset of 0 were chosen. Excitation and recording of emission with lasers were performed successively to avoid falsification of the fluorescence signal. 5 z-stack images were recorded of each biofilm on evenly distributed points of the anode. If the 10x objective was used, the spacing between recorded pictures was set to 3 μm; if the 16x objective was used, the spacing between recorded pictures was set to 0.925 μm.

Following measurements, CLSM images were analysed using the software NIS elements AR (Nikon, Düsseldorf, Germany). Fluorescence data of the biofilm was first binarised using a threshold intensity between 321 and 4,095 for the AO signal and a threshold intensity between 880 and 4,095 for the PI signal. From the binarised image, the total volume of biofilms (V_B) as well as the volume of nonviable cells (V_d) was determined. Binarised images were further used to determine the surface coverage (*co*) of the anodes. From these data, the biofilm thickness (*d*) was calculated by equation 2.4, where *a* represents the area of the recorded image (a = 1,612,290 μm² for 10x and a = 637,954 μm² for 16x magnification).

$$d = \frac{V_B}{co \cdot a} \tag{2.4}$$

The viability of biofilms (*L*) was determined from the total biofilm volume and the volume of PI stained cells, which were considered to be nonviable, by equation 2.5.

$$L = \frac{V_B - V_d}{V_B} \tag{2.5}$$

The biofilm length in the z-direction is distorted when using the 10x air objective due to the refractive index of water (Hell et al. 1993). In the case of the 16x water-dipping objective, such a distortion is not occurring because the objective is located within the same liquid as the biofilm (Hell et al. 1993). Therefore, a conversion factor was determined from images of the same biofilm recorded with 10x and 16x magnification. The thicknesses determined from biofilms differed by a factor of 0.672 ± 0.008. This factor was used to convert biofilm thickness measured at 10x (d_{10x}) magnification to the actual biofilm thickness (d_{16x}) by equation 2.6.

$$d_{10x} \cdot 0.672 = d_{16x} \tag{2.6}$$

2.7 Microaerobic cultivation in bioreactor system

G. sulfurreducens was cultivated non-electrochemically in a bioreactor system (DASGIP®, Eppendorf AG, Hamburg, Germany) to investigate the capability of this organism to reduce oxygen. Bioreactors consisted of glass vessels (total volume 1.1 L) equipped with 6-blade Rushton turbines (diameter 30 mm), a temperature sensor, a sampling and inoculation port, an optical dissolved oxygen (DO) sensor (Hamilton® Visiferm™ DO225, Hamilton Company, Reno, Nevada, USA) and a pH electrode (405-DPAS-SC-K8S/225, Mettler-Toledo, Columbus, Ohio, USA). Gassing was supplied via a gas sparger and exhaust gas was passed through a gas cooling system to minimise loss of fluid due to evaporation. Reactors were operated with a working volume of 500 mL.

Medium was prepared directly in reactor vessels according to **Table 2.12** and **Table 2.13**, omitting cysteine, vitamin and trace element solution, and 100 mL for the inoculum. Bioreactors were completely assembled and sterilised at 121 °C and 2 bar for 20 min. Subsequently, bioreactors were connected to the control unit and parameters for the cultivations were set to the following values: temperature: 30 °C, gas velocity: 6 sL/h, gas composition: air (21 % O_2), stirring speed: 200 min^{-1}. Cysteine, vitamin and trace element solution were added through the sampling and inoculation port using a sterile inoculation flask. Bioreactors were allowed to equilibrate overnight before calibration of DO sensors. DO sensor calibration was carried out at 21 % O_2 (DO signal of 100 %) and at 0 % O_2 (DO signal of 0 %), the latter corresponded to gassing with 100 % N_2. Following DO sensor calibration, bioreactors were inoculated using 100 mL of an overnight preculture of *G. sulfurreducens*.

Samples were taken at intervals of 2 h using 5 mL and 2 mL syringes. 5 mL of reactor suspension were removed first from the sampling port and discarded in order to completely rid the sampling port of old solution. Subsequently, 1 mL of sample was taken using a 2 mL syringe. The OD_{600} of samples was determined. Samples were centrifuged at 13,000 g and 4 °C for 3 min. The supernatant was filtered into an Eppendorf tube using cellulose-acetate or regenerated cellulose filters (pore size 0.2 µm) and samples were stored at -20 °C until HPLC analysis.

OD_{600} was used to calculate the biomass concentration using a conversion factor of 0.62 g_{CDW}/L (cell dry weight, CDW). This conversion factor was determined by gravimetric analysis. 5 mL to 10 mL of a *G. sulfurreducens* pure culture grown anaerobically with 40 mM of fumarate were filtered using cellulose-nitrate filters (Sartorius AG, Göttingen, Germany) and dried at 80 °C for 24 h. The amount of CDW was determined from the difference in weight of dried filters before and after biomass filtration and used to calculate the biomass concentration. Biomass concentration was plotted against the corresponding OD_{600} and the conversion factor was determined to be 0.62 ± 0.04 g/L by linear regression.

At the end of cultivations, bioreactors were disconnected from the control unit and transferred to the biosafety cabinet. This allowed for sterile sampling of a small volume of cultivation suspension directly from the bioreactor which was used to conduct a contamination test as described above.

6 different cultivations were carried out in the DASGIP® reactor system. The cultivations differed in the amount of electron acceptor (fumarate and oxygen) supplied. An overview of the experiments conducted is listed in **Table 2.14**. It was shown by Lin et al. (2004) that cell growth of *G. sulfurreducens* using oxygen as terminal electron acceptor is only possible if cells are primarily grown on fumarate. Therefore, all cultivations were started anaerobically with the gas composition set to 0 % O_2. Gas composition was changed in microaerobic experiments after 2 h of cultivation according to **Table 2.14**. Hereby, air and N_2 were mixed to obtain the indicated volumetric oxygen concentration. In experiment 4FSI, the gas composition was increased stepwise (SI) every 2 h depending on the biomass concentration determined by OD_{600} measurement.

Table 2.14: Experiment abbreviation and electron acceptor concentrations of experiments comparing anaerobic and microaerobic growth of *G. sulfurreducens*.

Experiment abbreviation	Fumarate [mM]	Oxygen in gas inlet [%]
40F0%	40	0
4F0%	4	0
4F1%	4	1
4F3%	4	3
4F5%	4	5
4FSI	4	Increasing from 3 to 12.5

2.8 Determination of volumetric mass transfer coefficient and maximum dissolved oxygen concentration

To quantify the amount of oxygen transferred into the medium during microaerobic cultivations, the volumetric mass transfer coefficient (k_La) and the maximum dissolved oxygen concentration (c_{O2}^*) were determined.

k_La was determined using the dynamic gassing out method. At this, oxygen was stripped from reactors using N_2 gas. The response of the DO sensor was monitored during the subsequent gassing with air (21 % O_2), 5 % O_2 or 3 % O_2. From the increase in the DO sensor signal, k_La was determined by linear regression according to equation 2.7 where c_{O2} is the DO concentration determined at timepoint t of the measurement.

$$k_La = \frac{\ln(c^*_{O2} - c_{O2})}{t} \quad (2.7)$$

k_La values were similar at 3, 5 and 21 % O_2 and the average k_La was determined to be $9.4 \pm 1.1\ h^{-1}$.

The maximum amount of oxygen soluble in the medium is strongly dependent on the concentration of ions from the salts used for the medium. Therefore, the maximum dissolved oxygen concentration (c_{O2}^*) was determined for the SW-minimal medium according to a method described by Weisenberger and Schumpe (1996). The salt solution as well as sodium from Na-acetate and Na-fumarate solution were regarded for calculations. The remaining solutions were disregarded due to the marginal salt concentrations and therefore negligible effects on the solubility of oxygen. c_{O2}^* was determined to be 0.31 mg_{O2}/L per % of oxygen in the gas composition.

2.9 Microarray based transcriptome analysis

To investigate the reaction of *G. sulfurreducens* to the presence of oxygen, transcriptome analysis was performed with samples taken from microaerobic cultivations 4F1%, 4F3% and 4F5%, comparing the transcript levels to cultivation 40F0%. Samples for transcriptomic RNA analysis were taken during the late growth phase of cultivations. 50 mL and 15 mL were sampled from each reactor in falcon tubes and centrifuged at 11,800 g and 4 °C for 10 min. The supernatant was discarded and cells were frozen in liquid nitrogen and stored at -80 °C until further treatment.

RNA was purified from cell pellets using a similar protocol as described previously (Biedendieck et al. 2011). Pellets were resuspended in 15 mg/mL lysozyme solution in TE-buffer (10 mM Tris HCl, 1 mM EDTA, pH 8). Cells were vortexed every 2 min for 10 s for a period of 30 min. Afterwards, cells were additionally disrupted mechanically with glass beads (diameter 150 µm to 212 µm) by vortexing for 3 min. Glass beads and cell fragments were removed by centrifugation at 10,000 g for 3 min. The RNeasy Mini Kit (Quiagen, Hilden, Germany) was used to purify RNA from the supernatant. The kit was used according to the manufacturer's instructions. Removal of chromosomal DNA was performed with DNAse (Quiagen, Hilden, Germany) and was performed twice to ensure complete removal. The concentration of RNA was measured with the NanoDrop ND-1000 (Peqlab Biotechnology, Erlangen, Germany). RNA quality was determined using the 2100 Bioanalyzer (Agilent Technologies, Santa Clara, California, USA) and an RNA Nano Chip (Agilent Technologies, Santa Clara, California, USA) following the manufacturer's instructions. Transcriptome analysis was only performed if samples had an RNA integrity number higher than 7.

Cyanine 3 and cyanine 5 from the ULS Fluorescent Labeling Kit for Agilent arrays from Kreatech (Leica Biosystems, Amsterdam, Netherlands) were used to stain purified RNA according to the manufacturer's instructions. Following staining, it was ascertained with the NanoDrop ND-1000 that the degree of labelling ranged between 1 and 3.6. The gene expression hybridization kit from Agilent (Agilent Technologies, Santa Clara California, USA) was used according to the manufacturer's instructions to fragment and hybridise RNA to a custom-made GE 8x15K microarray (Design ID 085375, Agilent Technologies, Santa Clara, California, USA) that was kindly designed by R. Biedendieck (Institute of Microbiology, Technische Universität Braunschweig, Germany). Following this, microarrays were analysed with the Microarray C Scanner (Agilent Technologies, Santa Clara, California, USA) as described previously (Borgmeier et al. 2011). Raw data was kindly evaluated by R. Biedendieck. Genes displaying a fold change (FC) $|\log_2 FC|$ of > 0.1 were considered to be differentially expressed. If of interest, genes displaying a $|\log_2 FC|$ of > 0.08 were also considered.

2.10 High performance liquid chromatography methodology

The concentration of organic acids was determined from samples by high performance liquid chromatography (HPLC). An HPLC system (institute's denotation: C2) from LaChrom Elite® (Hitachi, Tokyo Japan) was used with the stationary phase consisting of an Aminex® HPX-87H column (Bio-Rad Laboratories, Hercules, California, USA) and the mobile phase consisting of 12.5 mM H_2SO_4. Depending on the sample, two different methods were used to assess organic acid concentrations. Samples containing both fumarate and acetate were determined at 25 °C with a flow rate of 0.3 mL/min for 40 min. Samples containing both succinate and lactate were determined at 45 °C with a flow rate of 0.5 mL/min for 23 min. Correspondingly, two different organic acid standards were measured as well for quantification, one containing 2-oxoglutarate (20 min), pyruvate (21 min), succinate (27 min), acetate (33 min) and fumarate (36 min) (institute's denotation: AF2.0 from 10/03/2015, retention times indicated in brackets) and the second containing oxalic acid (9 min), citric acid (10 min), 2-oxoglutarate (11 min), pyruvate (13 min), succinate (15 min), lactate (16 min), formic acid (17 min), acetate (19 min), propionate (22 min) and butyrate (27 min) (institute's denotation: oS from 06/03/2017, retention times indicated in brackets). Measurements were performed and evaluated using the OpenLab software (Agilent Technologies, Santa Clara, California, USA). For detection of substances, a refractive index detector as well as a diode array detector (210 nm or 230 nm for fumarate) were used. Measurements of the organic acid standards were used to determine the conversion factors between detected peak area and concentration by linear regression and these factors were used to determine the concentration of organic acids in samples.

The determined substrate concentrations were used to calculate the yield coefficients of biomass produced per acetate ($Y_{X/S}$), fumarate reduced per acetate ($Y_{Fum/Ac}$), oxygen reduced per acetate ($Y_{O2/Ac}$) and succinate produced per fumarate ($Y_{Suc/Fum}$) by equations 2.8 to 2.11, where X, c_{Ac}, c_{Fum}, c_{O2} and c_{Suc} are the mass or molar concentrations of biomass, acetate, fumarate, oxygen and succinate, respectively.

$$Y_{X/S} = \frac{X}{c_{Ac}} \tag{2.8}$$

$$Y_{Fum/Ac} = \frac{c_{Fum}}{c_{Ac}} \tag{2.9}$$

$$Y_{O2/Ac} = \frac{c_{O2}}{c_{Ac}} \tag{2.10}$$

$$Y_{Suc/Fum} = \frac{c_{Suc}}{c_{Fum}} \tag{2.11}$$

3 Results and Discussion

3.1 Mixed and pure cultures in microbial electrolysis cell

3.1.1 Current production of pure cultures and mixed culture

Geobacter sulfurreducens and *Shewanella oneidensis* were each cultivated in a microbial electrolysis cell (MEC) setup as pure cultures at an anode potential of 0.2 $V_{Ag/AgCl}$ with 10 mM of acetate or lactate, respectively. The resulting chronoamperograms of cultivation cycle 1 of *G. sulfurreducens* can be seen in **Figure 3.1**. The onset time for current generation was at 6 h after inoculation. The onset in current generation can be explained by the attachment of some *G. sulfurreducens* cells onto the graphite anode and an initial oxidation of acetate and reduction of the electrode for energy generation. The subsequent exponential increase in current suggests an exponential cell growth of *G. sulfurreducens* during the initial biofilm formation. This was followed by a linear increase in current, probably concomitant with a linear increase in biofilm thickness, until a maximum current density of 0.39 ± 0.09 mA/cm^2 was achieved after 4 days. The maximum current density achieved with *G. sulfurreducens* is in agreement with observations by others (Marsili et al. 2008b; Stephen et al. 2014). For instance, Zhu, Yates and Logan (2012) saw about 0.57 mA/cm^2 on graphite electrodes set to 0.3 $V_{Ag/AgCl}$ with 10 mM of acetate in a reactor designed very similar to the system used here. At the point of peak current, the acetate level within the cultivation medium was likely too low to allow for the maintenance of current. Therefore, the current density declined again to 0.035 mA/cm^2.

No sample was taken for acetate determination at the time of peak current to avoid disturbance of the cultivation e.g. by intruding oxygen. But from a sample taken at the end of the cultivation it could be seen that acetate was consumed almost completely (residual acetate < 0.1 mM). From the acetate consumed and the charge transferred in total, the Coulombic efficiency (CE) can be calculated to 105 ± 5 %. This value appears unrealistic at first glance since it should not be possible to find more electrons in the form of electrical current than there are present in the provided electron donor. However, a reasonable explanation is the recycling of H_2. As an airtight, single chamber system was used for cultivation, part of the H_2 formed at the cathode remained dissolved in the cultivation medium. *G. sulfurreducens* can use H_2 as electron donor (Caccavo *et al.* 1994). Therefore, it is possible that *G. sulfurreducens* consumed H_2 and transferred electrons to the anode a second time, thus erroneously increasing the resulting CE. H_2 recycling has previously been reported by others (Call et al. 2009; Liu et al. 2015) and also explains why current density evens out at 0.035 mA/cm^2 instead of decreasing to 0 mA/cm^2.

Beyond that, the large CE seen for *G. sulfurreducens* is an indicator for metabolically active and viable cells. Similarly high CEs of 95 % have been reported by Bond and Lovley (2003) in a two-chamber system.

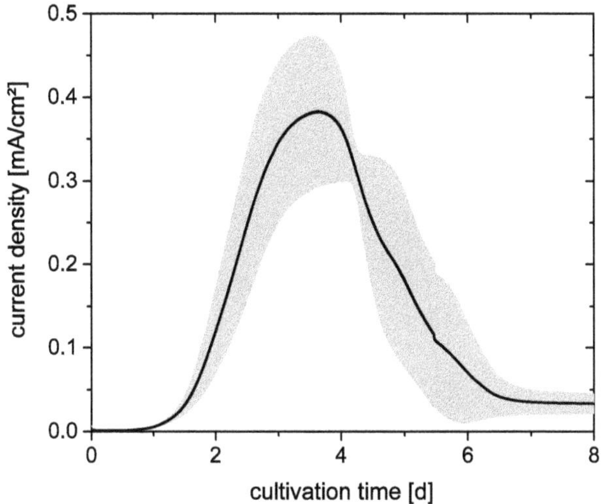

Figure 3.1: Chronoamperogram of *G. sulfurreducens* pure culture (cycle 1) grown at 30 °C on graphite electrodes poised to 0.2 $V_{Ag/AgCl}$ with 10 mM of acetate. Grey area indicates standard deviation from biological triplicates.

S. oneidensis began to produce current within the first hour after inoculation (see **Figure 3.2**) (data from Upmann, 2018). This faster initial current generation compared to *G. sulfurreducens* can be explained by the indirect flavin-based form of electron transfer that *S. oneidensis* is capable of. Riboflavin was present in the medium used so that current generation was immediately possible even without the need of *S. oneidensis* to first produce this shuttle molecule. Current increased, however not in an exponential manner, up to 0.0017 ± 0.0005 mA/cm^2. This value is more than 200-fold lower compared to the maximum current density achieved with *G. sulfurreducens* and demonstrates the better suitability of *G. sulfurreducens* as a current producing organism. In the literature, current densities of 0.016 to 0.02 mA/cm^2 have been reported for *S. oneidensis* (Marsili et al. 2008a; Rosenbaum et al. 2010; Rosenbaum et al. 2011). Those are about 10-fold higher as current densities observed here. This indicates non-optimal growth conditions for *S. oneidensis* in the MEC setup used here.

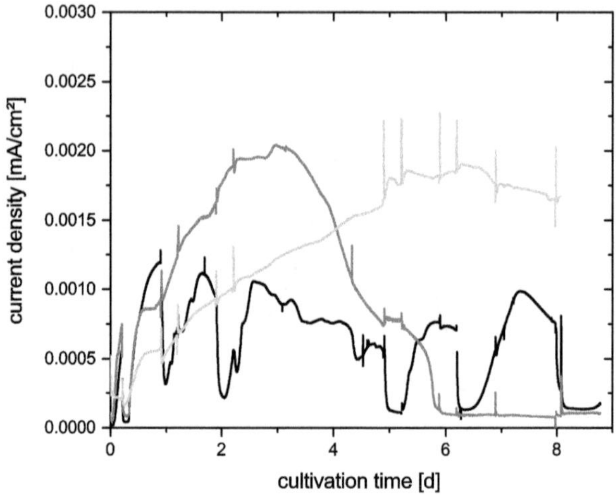

Figure 3.2: Chronoamperograms of *S. oneidensis* pure cultures grown at 30 °C on graphite anodes poised to 0.2 $V_{Ag/AgCl}$ with 10 mM of lactate. Differently coloured lines represent individual replicates. Spikes are due to sampling. (Upmann 2018)

Samples were taken readily to investigate the course of substrate consumption. The substrate lactate was consumed and degraded to acetate (48 ± 11 %) and pyruvate (23 ± 4 %), but the absolute lactate consumption differed between the three replicates. Complete consumption was only achieved in one replicate while in the other two replicates lactate consumption ceased after 2 and 6 days, leaving a residual concentration of 6.2 mM and 4.7 mM, respectively. *S. oneidensis* is not able to completely oxidise lactate under anaerobic conditions, which is the reason why acetate accumulates (Pinchuk *et al.*, 2011). The CE achieved was below 1 % in all three replicates. This is a further indicator that *S. oneidensis* was not capable of using the anode properly as electron acceptor. In this case, *S. oneidensis* needs to dispose of electrons from NADH in a different way. Succinate was produced and excreted in all experiments at concentrations of up to 0.53 mM (see appendix (chapter 6), Figure A.9), which indicates a reverse carbon flow into the TCA cycle. Formation of succinate requires NADH, thus NADH produced during lactate consumption can be reoxidised to NAD^+. Another possibility is the production of H_2 from lactate, which was shown for *S. oneidensis* by Meshulam-Simon et al. (2007). It is also possible that small amounts of oxygen leaked into the reactors during the sampling process. Slightly different amounts of oxygen intruding in each of the three replicates could explain the high irreproducibility seen both in the substrate levels and in the current production.

G. sulfurreducens and *S. oneidensis* were also cultivated together in the MEC setup as a defined mixed culture. The process conditions applied were the same as for the pure cultures except for the provision of 5 mM acetate and 5 mM lactate instead of 10 mM of either substrate as electron donor. Chronoamperograms of cultivation cycle 1 are depicted in **Figure 3.3 A**. Individual replicates are shown here as the experiments did not yield reproducible results. Two different trends could be observed. The first was similar to the *G. sulfurreducens* pure culture with the current density increasing within four days (black line in **Figure 3.3 A**). The second showed a slowly increasing current and the maximum value was not reached until day 10 (blue lines in **Figure 3.3 A**). Yet, the maximum current density seen was 0.54 ± 0.07 mA/cm^2 which is 38 % higher compared to the *G. sulfurreducens* pure culture. This indicates a positive effect resulting from the combined cultivation. In similar experiments, Prokhorova et al. (2017) saw improved maximum current densities in a defined mixed culture of *G. sulfurreducens*, *S. oneidensis* and *Geobacter metallireducens* (0.62 mA/cm^2) compared to a *G. sulfurreducens* pure culture (0.47 mA/cm^2). This increase of 32 % in the study of Prokhorova et al. (2017) corresponds well to the increased maximum current density seen here.

Samples were taken readily to observe the concentration profiles of the two substrates acetate and lactate, which are shown in **Figure 3.3 B**. As was to be expected from the unreproducible current profiles, there were also different concentration profiles seen for the substrates. But the general trend was similar in all replicates, whereby lactate was degraded first, resulting in a slight increase of acetate. From day 2 respectively day 6 onwards, acetate was degraded as well. Acetate degradation is linked to current production. An increase in current starts when the highest acetate concentration is detected and the highest current is measured when the highest rate of acetate degradation is occurring (see **Figure 3.3**). This suggests *G. sulfurreducens* to be responsible for the main current production since only *G. sulfurreducens* can degrade acetate under anaerobic conditions. The CE determined for this experiment was 121 ± 13 % which is in the same range as for the *G. sulfurreducens* pure culture and indicates metabolically active and viable cells.

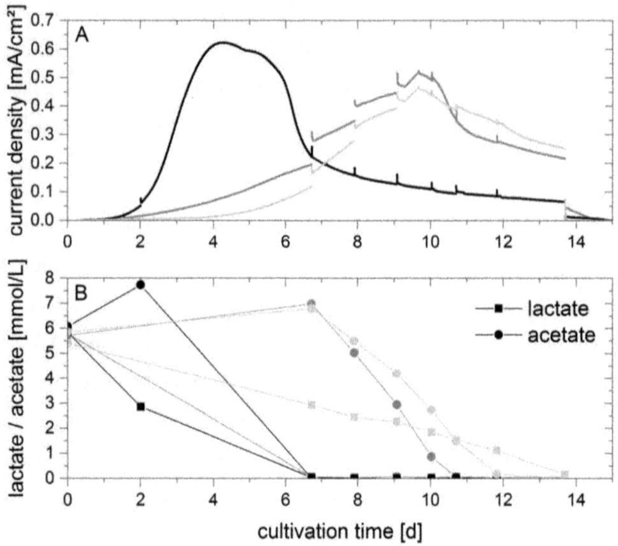

Figure 3.3: (A) Chronoamperograms and (B) substrate concentrations of cultivation cycle 1 of defined mixed cultures of *G. sulfurreducens* and *S. oneidensis* grown at 30 °C on graphite electrodes poised to 0.2 $V_{Ag/AgCl}$ with 5 mM of acetate and 5 mM of lactate. Differently coloured lines represent individual data from biological triplicates. Spikes in (A) are due to sampling.

The question remains why different substrate consumption profiles are observed when the same electrochemical setup and cells from the same precultures were used for experiments. One possible explanation for this phenomenon is lactate consumption directly by *G. sulfurreducens*, which does typically not occur. It was shown by Summers et al. (2012) that the ability for lactate degradation is caused by a single base pair mutation in a transcriptional regulator encoded by GSU0514. The unpredictability of such a mutation at a random time point during experiments can explain the differences in lactate consumption and the resulting differences in acetate and current profiles.

Current production from lactate has been reported previously for *G. sulfurreducens* pure cultures (Call and Logan 2011; Speers and Reguera 2012). The thereby proposed flow of carbon from lactate is shown in **Figure 3.4**. Lactate is degraded to pyruvate and subsequently acetyl-CoA (Speers and Reguera 2012). Acetyl-CoA would usually be oxidised to CO_2 in the TCA cycle. But an essential enzyme of the TCA cycle – the succinyl-CoA-synthetase which converts succinyl-CoA to succinate – is not active in *G. sulfurreducens* (Segura et al. 2008). Thus, the

TCA cycle cannot function completely, and acetyl-CoA oxidation is prevented. Absence of succinyl-CoA-synthetase activity is not problematic for the usually occuring acetate degradation. Acetate is metabolised to acetyl-CoA via an acetyl-CoA-transferase which transfers the CoA-residue from succinyl-CoA to acetate (see grey arrows in **Figure 3.4**) (Segura et al. 2008). Thereby succinate is formed from succinyl-CoA, which eliminates the need for succinyl-CoA-synthetase activity (Segura et al. 2008). With lactate as substrate, however, acetyl-CoA accumulates. To dispose of this metabolic glut, carbon flux is directed from acetyl-CoA to acetate which is then secreted (Segura et al. 2008). This is in accordance with an accumulation of acetate following lactate degradation seen here. Similarly, Speers and Reguera (2012) found an accumulation of acetate as a result of lactate degradation in electrochemical pure cultures of *G. sulfurreducens*. The spontaneous mutation in the transcriptional regulater encoded by GSU0514 activates the transcription of the succinyl-CoA synthetase (see orange arrow in **Figure 3.4**) (Summers et al. 2012). Thus, the TCA can function completely, lowering the metabolic glut generated by lactate and enabling faster lactate consumption. The substrate profiles obtained here suggest that a mutation in GSU0514 occurred at an early stage of cultivation in the experiment represented by the black line (see **Figure 3.3**) but only at a later stage in the other two replicates.

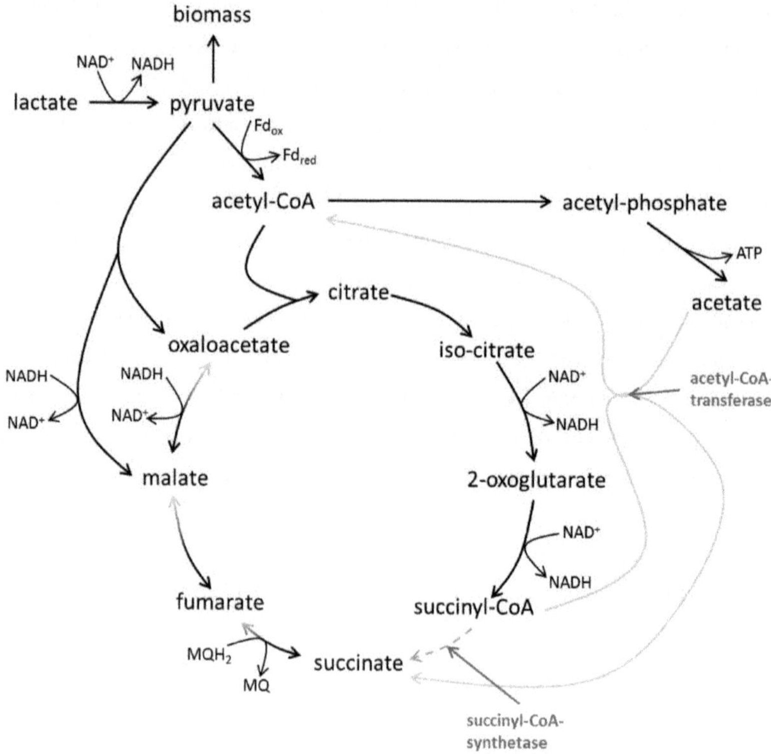

Figure 3.4: Proposed flow of carbon (black arrows) in *G. sulfurreducens* from lactate consumption as derived from Speers and Reguera (2012) and Segura et al. (2008). Grey arrows indicate reactions normally occurring during acetate degradation with reduced carbon flow when lactate is present. Orange arrow indicates reaction only possible after single base pair mutation in GSU0514. Fd = ferrodoxin, MQ = menaquinone, MQH_2 = menaquinol. Red annotations refer to enzymes responsible for the indicated reactions.

Current production was mainly observed in response to acetate degradation and not to lactate degradation. HPLC-analysis revealed an accumulation of succinate (see appendix (chapter 6), Figure A.12). This corresponds to the observation made with *S. oneidensis* pure cultures, although here succinate accumulated to a greater extend as was seen there (approx. 1.8 mM to 2.5 mM vs. 0.16 mM to 0.53 mM). Succinate can be derived from pyruvate via anaplerotic reactions (see **Figure 3.4**). Hereby, NADH produced during lactate oxidation is consumed. Electrons from NADH used to produce succinate are no longer available for current production, which explains the low current generation seen during the initial lactate degradation. Additional experiments are necessary to distinguish whether *S. oneidensis* is responsible for the enhanced succinate production caused by the presence of another bacterial species, or whether this accumulation is in part due to *G. sulfurreducens*.

It could be assumed that only *G. sulfurreducens* is responsible for the observed lactate degradation. However, in the work of Speers and Reguera (2012) and Call and Logan (2011), current generation from lactate in pure cultures was poorer compared to current production from acetate. Here, however, an increase in current generation due to the use of lactate in a combined cultivation with *S. oneidensis* was observed. This indicates that *S. oneidensis* is in part responsible for lactate degradation and for the enhanced current production by *G. sulfurreducens*.

3.1.2 Long-term behaviour of defined mixed culture

The combined cultivation of *G. sulfurreducens* with other organisms has previously been shown to be beneficial, for example by Dolch et al. (2014) and Prokhorova et al. (2017), as was presented in chapter 1.6.1. However, these authors only observed short-term effects as experiments lasted a maximum of 7 days. An integration of *S. oneidensis* into the *G. sulfurreducens* based biofilm could be observed, but it was not investigated whether this integration was stable in the long term, especially if planktonic cells are removed from the surrounding medium.

Therefore, a second and third cultivation cycle were performed here with the pure culture of *G. sulfurreducens* and the defined mixed culture, where medium was exchanged after current depletion to select for sessile cells. The chronoamperograms of cultivation cycle 2 of *G. sulfurreducens* can be seen in **Figure 3.5** including cycle 1 for comparison. Current increased immediately after medium exchange as the electrochemically active biofilm was already present on the anode as opposed to cycle 1 where *G. sulfurreducens* had to colonise the electrode first. Maximum current densities achieved in cycle 2 and 3 were similar compared to cycle 1 with 0.40 ± 0.05 mA/cm^2 and 0.35 ± 0.06 mA/cm^2, respectively. The CE was higher compared to cycle 1, reaching 136 ± 16 %. The higher CE compared to cycle 1 together with the consistent maximum current density are an indicator that *G. sulfurreducens* did not continue growing extensively in cycle 2 but rather efficiently turned acetate into current.

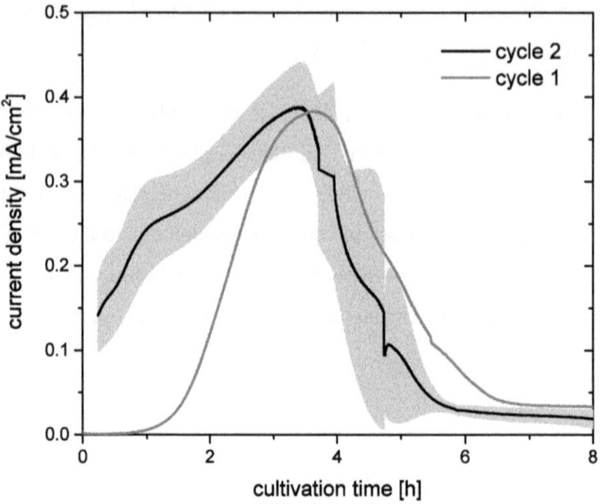

Figure 3.5: Chronoamperogram of *G. sulfurreducens* pure culture (cycle 2, black line) grown at 30 °C on graphite electrodes poised to 0.2 $V_{Ag/AgCl}$ with 10 mM of acetate. Grey area indicates standard deviation from biological triplicates. Blue line represents chronoamperograms from cycle 1 (see **Figure 3.1**) for comparison.

Chronoamperograms as well as acetate and lactate concentrations of cycle 2 of the defined mixed culture are shown in **Figure 3.6**. In plain contrast to cycle 1 (see chapter 3.1.1), chronoamperograms and substrate consumption in cycle 2 were highly similar between replicates. Differences in lactate consumption in cycle 1 were hypothesised above to be caused by the unreproducibility of a spontaneous mutation enabling *G. sulfurreducens* to consume lactate. This hypothesis is reinforced by the fact that there were no differences in lactate degradation in cycle 2, indicating that the mutation happened in all replicates.

As opposed to lactate degradation prior to acetate, substrates were consumed in reverse order, creating two distinct maxima in the current profile. Acetate degradation and current production started immediately similar to observations made in the *G. sulfurreducens* pure culture, which suggests that an established biofilm can reduce acetate and transfer electrons faster. Efficient transfer of electrons to the anode also renders the disposal of NADH produced during lactate degradation unnecessary, which in cycle 1 was linked to increasing succinate concentrations. Succinate determined during cycle 2 supports this as concentrations did not increase more than 0.1 mM.

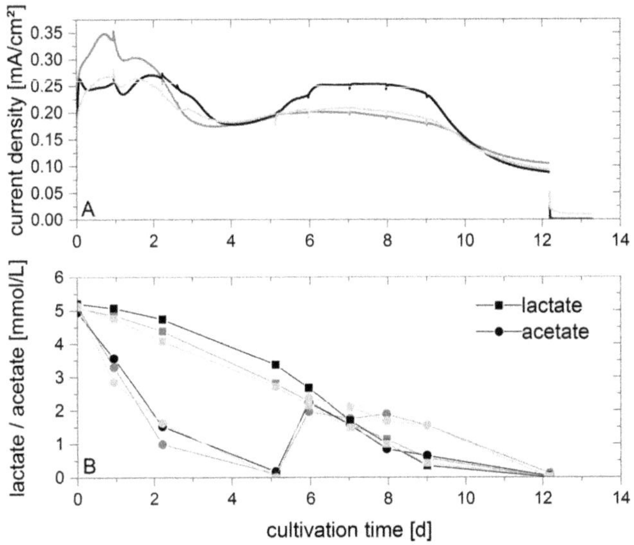

Figure 3.6: (A) Chronoamperograms and (B) substrate concentrations of cultivation cycle 2 of defined mixed cultures of *G. sulfurreducens* and *S. oneidensis* grown at 30 °C on graphite electrodes poised to 0.2 $V_{Ag/AgCl}$ with 5 mM of acetate and 5 mM of lactate. Differently coloured lines represent individual data from biological triplicates.

Overall, the maximum current density achieved in cycle 2 and 3 was lower than in cycle 1, only reaching 0.30 ± 0.04 mA/cm^2 and 0.24 ± 0.05 mA/cm^2, respectively. This differs from observations made with the *G. sulfurreducens* pure culture. Thus, it appears that the positive effect of *S. oneidensis* seen in cycle 1 is not continued after medium replacement. Planktonic *S. oneidensis* cells were removed during medium replacement to select for sessile *S. oneidensis* cells. The poorer performance of the defined mixed culture in cycles 2 and 3 might be an indicator that the amount of *S. oneidensis* potentially located in the biofilm is too low to maintain the positive effect. To test this, samples of biofilm and planktonic cells were taken during cycle 3 for flow cytometry analysis to quantify *S. oneidensis* both in the biofilm and planktonically.

Flow cytometry is a useful technique to determine the distribution of different bacterial species and was used before to analyse electrochemically active cells (Müller 2007; Koch et al. 2014). Measurements were carried out in collaboration with the Helmholtz Centre for Environmental Research (UFZ) in Leipzig, Germany, to determine the distribution of *S. oneidensis* and *G. sulfurreducens* in the defined mixed culture. Representative dot plots from both cell types in pure culture as well as in the defined mixed culture are shown in **Figure 3.7** for planktonic and biofilm descendent cells. Dot plots show the DAPI fluorescence signal (y-axis) and the scattered light (forward scatter, x-axis) of individual cells. Colours indicate the amount of cells found at a specific spot, red meaning a high amount of cells and blue a low amount. Black lines indicate the cell gates in which to expect *S. oneidensis* or *G. sulfurreducens* cells and which were established by measurements of pure cultures. However, a few cells in pure culture samples are scattering into the other species' cell gate, creating a falsely positive signal. This falsely positive signal was 1 % in the case of a *G. sulfurreducens* pure culture biofilm. In the defined mixed culture biofilm, *S. oneidensis* was only detected to about 2 %, which is not significantly higher than the falsely positive signal. Therefore, it is possible there were no *S. oneidensis* cells incorporated into the biofilm. Rather, the anodic biofilm was dominated by *G. sulfurreducens*.

When Dolch et al. (2014) analysed the amounts of sessile cells in a *G. sulfurreducens* / *S.* oneidensis mixed culture and in respective pure cultures, they found greater cell numbers of both species in biofilms on graphite felt anodes. The amount of *G. sulfurreducens* was increased 1.7-fold in the mixed culture compared to the pure culture, and the number of sessile *S. oneidensis* cells was even increased 11-fold (Dolch et al. 2014). The distribution of both species within the biofilm was 91 % for *G. sulfurreduens* and 9 % for *S. oneidensis* (Dolch et al. 2014). Thus, these authors could detect *S. oneidensis* in the biofilm in contrast to the work presented here. But experiments by Dolch et al. (2014) were only performed for less than two days, hence the incorporation of *S. oneidensis* was not investigated in terms of long-term stability.

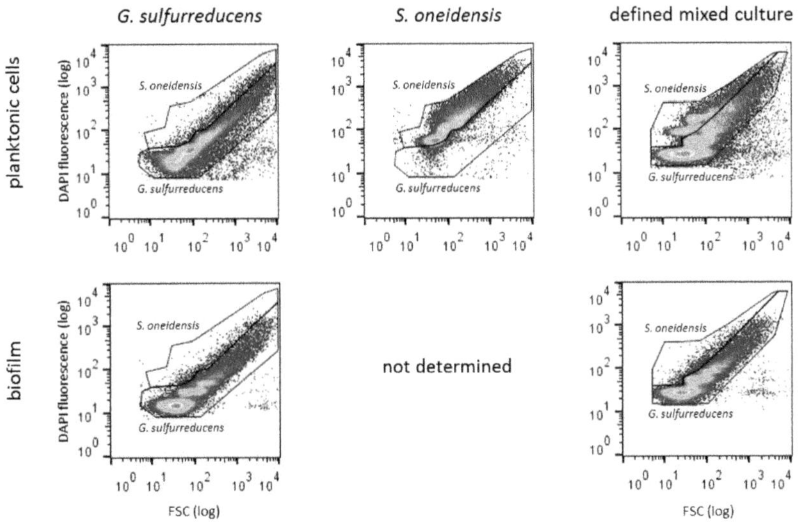

Figure 3.7: Representative dot plots from planktonic and biofilm derived cells from a *G. sulfurreducens* pure culture, an *S. oneidensis* pure culture (only planktonic) and the defined mixed culture. Colours indicate cell counts (red = high number of cell counts, blue = low number of cell counts).

Slightly longer time periods were investigated by Prokhorova et al. (2017), who studied mixed cultures of three bacteria, namely *G. sulfurreducens*, *G. metallireducens* and *S. oneidensis*, to determine whether *G. sulfurreducens* ultimately overgrows *S. oneidensis*. Activated carbon cloth was set to a potential of -0.2 $V_{Ag/AgCl}$ and cell numbers for all three bacteria were determined during 7 days of cultivation (Prokhorova et al. 2017). *S. oneidensis* and *G. metallireducens* were found to be capable of growing inside the biofilm (Prokhorova et al. 2017). Still, *G. sulfurreducens* was dominant in the biofilm (approx. 95 %) compared to *S. oneidensis* (approx. 5 %) and *G. metallireducens* (< 0.1 %) (Prokhorova et al. 2017). The low abundance of cells in the biofilm seen here (after several weeks) compared to Prokhorova et al. (2017) indicates that *S. oneidensis* is not incorporated stably in the biofilm. Rather, cells are being dislodged by *G. sulfurreducens*. Thus, the positive effect seen by Dolch et al. (2014) and Prokhorova et al. (2017) and also seen here in cycle 1 cannot be sustained over a longer period of time. It is of course also possible that *S. oneidensis* was not incorporated into the biofilm even within the first few days of experiments performed here because of a different MEC system and anode material used and that in the works performed by Dolch et al. (2014) and Prokhorova et al. (2017) the incorporation might have been stable for a longer time-period.

To investigate this, flow cytometry should also be performed after cycle 1 or even during different time points of the cycle, for example at the time of highest current production.

Another difference between the work performed here and the study of Prokhorova et al. (2017) is the choice of provided carbon source. While here both acetate and lactate were provided, Prokhorova et al. (2017) only used lactate as carbon source. The authors argue that the stability of their mixed culture lies in the fact that *G. sulfurreducens* and *G. metallireducens* are dependent on the metabolic products from lactate degradation by *S. oneidensis*, i.e. acetate and propionate, respectively. But from the observations made here it is more likely that *G. sulfurreducens* itself is at least in part responsible for lactate degradation and is thus not dependend on the presence of *S. oneidensis*. Even within the work of Prokhorova et al. (2017) evidence can be found that *G. sulfurreducens* performs lactate degradation. Transcriptome analysis from that study revealed a higher expression of the gene GSU1059 that encodes for the succinyl-CoA synthetase discussed above (see **Figure 3.4**) under mixed culture conditions, indicating that the mutation in the transcriptional regulator GSU0514 discovered by Summers et al. (2012) also took place in the experiments performed by Prokhorova et al. (2017). Thus, the dependency of *G. sulfurreducens* on lactate degradation by *S. oneidensis* might be obsolete even if only lactate were given as carbon source. Therefore, it should be investigated in future experiments how the defined mixed culture behaves if only lactate is given as carbon source and whether the presence of *S. oneidensis* can really be stabilised in this way.

The amount of planktonic *S. oneidensis* doubled within 7 days in the work of Prokhorova et al. (2017). Here, however, planktonic cells were removed twice to select for sessile cells. By the end of cultivation cycle 3, planktonic cells were still found in the medium of MECs, but only to an OD_{600} of 0.086 ± 0.013. Of these, only 25 % were *S. oneidensis* as determined by flow cytometry. This amount is much lower compared to the initial amount of planktonic *S. oneidensis* used for inoculation (OD_{600} of 0.1), which is another indicator that the positive effect on current generation seen here in cycle 1 and seen by Prokhorova et al. (2017) is based on planktonic cells of *S. oneidensis*.

An explanation for this positive effect of planktonic *S. oneidensis* cells could be the generation of H_2. It was shown by Meshulam-Simon et al. (2007) that *S. oneidensis* can produce H_2 from lactate when under electron acceptor limiting conditions. The anode used here was seen to be a subpar electron acceptor for *S. oneidensis* in pure culture, thus, H_2 production by *S. oneidensis* is likely to have occurred. In the defined mixed culture, H_2 produced by *S. oneidensis* in cultivation cycle 1 could have resulted in the higher current generation by *G. sulfurreducens*.

Correspondingly, Prokhorova et al. (2017) discovered two hydrogenase genes in *S. oneidensis* that are expressed at higher levels under mixed culture conditions. This affirms the above given explanation.

To establish a stable defined mixed culture, it is therefore necessary to ensure permanent presence of planktonic *S. oneidensis* cells. This could be realised with a continuously operated MEC reactor with a dilution rate below the maximal specific growth rate of *S. oneidensis*. Planktonic *S. oneidensis* cells are able to achieve a stationary state and thus a constant cell concentration in such a continuously operated system. A flow cell design, however, would not be advantageous as planktonic *S. oneidensis* cells as well as the flavins they produce would be washed out.

Another possibility of improving growth conditions for *S. oneidensis* is the controlled introduction of certain amounts of oxygen into the MEC. Although oxygen decreases Coulombic efficiencies in MECs, it has been shown to improve cell growth and current production by *S. oneidensis* (Rosenbaum et al. 2010). Thus, oxygen might stabilise the presence of *S. oneidensis* even within the biofilm.

If a long-term stable defined mixed culture of *G. sulfurreducens* and *S. oneidensis* could be established, this culture might be applied for H_2 production. Call et al. (2009) found H_2 production rates impaired in undefined mixed cultures due to the presence of methanogenic bacteria in the consortium which used part of the H_2 produced at the cathode to form methane. They showed that pure cultures of *G. sulfurreducens* are capable of producing H_2 at the same rate as undefined mixed cultures (Call et al. 2009). Seeing as *S. oneidensis* improves current generation by *G. sulfurreducens*, the defined mixed culture offers the advantages of positive interactions between species seen in mixed cultures. But because they are lacking the ability for methane production, no loss of H_2 should be expected from the defined mixed culture. Thus, H_2 recoveries are expected to be higher in the defined mixed culture used here compared to an undefined consortium containing methanogens.

3.1.3 Biofilm formation on anodes

Biofilms were analysed with confocal laser scanning microscopy (CLSM) after experiments were completed. From CLSM images the thickness of anodic biofilms was determined. *G. sulfurreducens* formed area-covering biofilms of 69 ± 18 μm thickness. This is in the normal range for *G. sulfurreducens* as thicknesses between 50 µm and several hundred micrometers have been reported in the literature (Babauta et al. 2012b; Leang et al. 2013; Renslow et al. 2013). *S. oneidensis* biofilms, on the other hand, only grew to 7 ± 4 µm. For comparison, 3D-pictures of both biofilms are shown in **Figure 3.8**. Cells were stained with two fluorescent dyes which allowed the calculation of the viability of biofilms. In both cases, viabilities were very high with 99 % and 100 % for *G. sulfurreducens* and *S. oneidensis*, respectively. Due to this, the low red fluorescence signal from nonviable cells is hardly distinguishable in **Figure 3.8**.

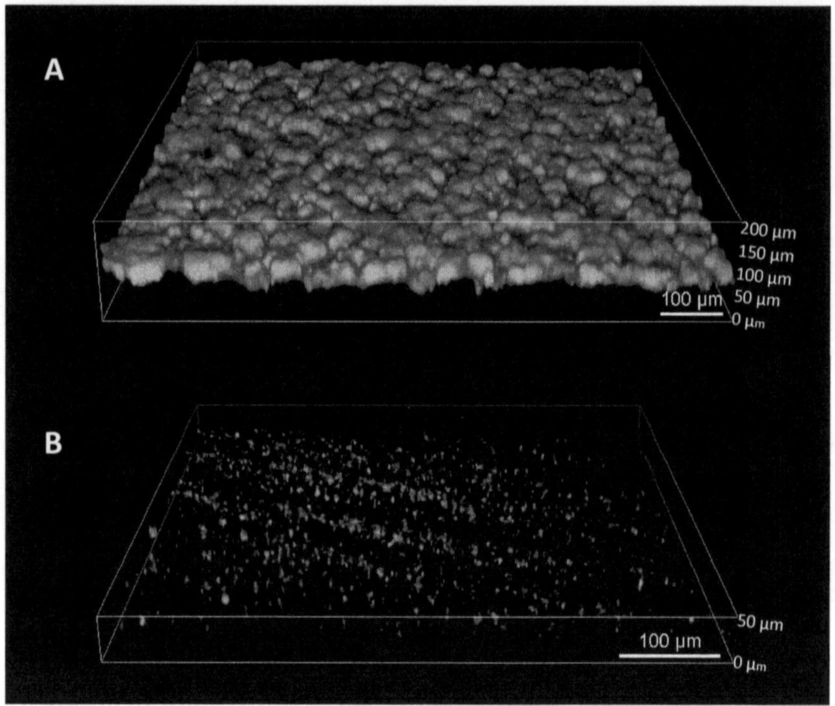

Figure 3.8: 3D-views from confocal laser scanning microscopy images of (A) *G. sulfurreducens* and (B) *S. oneidensis* on graphite anodes. Green represents acridine orange (cells with intact membrane), red represents propidium iodide (cells with impaired membrane).

The superior biofilm formation ability of *G. sulfurreducens* compared to *S. oneidensis* has also been reported previously (Dolch et al. 2014). The thin *S. oneidensis* biofilm does not allow for extensive direct electron transfer, which elucidates the poorer current production seen with this bacterium. Alternatively, *S. oneidensis* produces flavins and performs indirect electron transfer

(Marsili et al. 2008a). It has been shown that flavin mediated electron transfer constitutes 75 % in *S. oneidensis* (Kotloski and Gralnick 2013). Flavin production creates a continuous metabolic burden and thus lowers the productivity of *S. oneidensis* (Kotloski and Gralnick 2013). This connotes a disadvantage compared to *G. sulfurreducens* since electrons in the thick and conductive biofilms of the latter are transferred directly and fast (Malvankar and Lovley 2014).

CLSM analysis of the defined mixed culture revealed a biofilm of 93 ± 8 µm covering the entire anode (see **Figure 3.9**). Just like the pure cultures, mixed cultures also displayed a very high viability of 100 %. The determined thickness was about 34 % increased compared to the *G. sulfurreducens* pure culture, but it should be noted that the standard deviations of the determined thicknesses overlap and that this difference is therefore not statistically significant. Nevertheless, this increase is noticeably close to the 38 % increased maximum current density, which might suggest a correlation between current density and biofilm thickness. Such a correlation has previously been reported, for example by Zhu et al. (2014) and Baudler et al. (2015), who both saw a linear correlation between biomass respectively biofilm thickness and current density. In the experiments performend here, biofilm thickness was only observed after cultivation cycle 3 by which time the maximum current density had decreased significantly. Thus, one could argue that this comparison and therefore the correlation are incoherent. However, from further experiments it could be seen that the thickness of the mixed culture biofilms is already achieved after one cultivation cycle and can therefore be linked to the current density in cycle 1 (Block 2018).

The increased biofilm thickness corresponds well to the findings of Dolch et al. (2014) that the amount of *G. sulfurreducens* cells increases 1.7-fold in mixed cultures compared to pure cultures. Thus, the enhanced biofilm thickness can be related to the presence of *S. oneidensis*. A central factor for biofilm formation and stability in *S. oneidensis* is extracellular DNA (eDNA) (Gödeke et al. 2011; Binnenkade et al. 2014). This eDNA is released by cell lysis of a subpopulation of cells (Gödeke et al. 2011; Binnenkade et al. 2014). Possibly, eDNA release by *S. oneidensis* is also positive for the biofilm formation of *G. sulfurreducens*, leading to the observed higher biofilm thickness seen here.

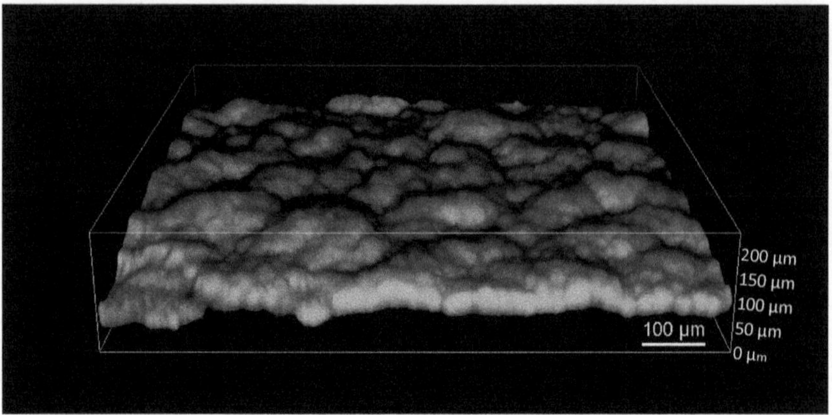

Figure 3.9: 3D-view from confocal laser scanning microscopy images of defined mixed culture of G. sulfurreducens *and* S. oneidensis *on graphite anodes. Green represents acridine orange (cells with intact membrane), red represents propidium iodide (cells with impaired membrane).*

A correlation between biofilm thickness and current density observed would suggest that the entire biofilm is metabolically active. Evidence for a complete metabolic activity has been found by several studies. Franks et al. (2010) dyed biofilms with 5-cyano-2,3-ditolyl tetrazolium chloride, which is a redox active molecule that can be reduced by metabolically active cells to the fluorescent formazan. A fluorescence signal was seen in the entire biofilm, indicating the general metabolic activity of all cells (Franks et al. 2010). Results from Stephen et al. (2014) correspondingly found equal amounts of acetate kinase independent of the distance to the electrode surface, indicating the same level of metabolic activity in all cells.

However, transcriptome analysis revealed that the expression of some genes differs depending on the distance to the electrode (Franks et al. 2010). Results suggested that cells within 20 µm of the electrode had higher growth rates compared to cells 30 µm to 60 µm above the electrode surface (Franks et al. 2010). Stephen et al. (2014) found OmcB to be present at higher concentrations at greater distances from the electrode. Energy used for OmcB production is no longer available for cell growth, which might explain the lower growth rates in upper biofilm layers seen by Franks et al. (2010). Greater abundance of OmcB allows some cytochromes to remain in a reduced state without reducing electron transfer rates (Liu and Bond 2012). Thus, a redox potential gradient necessary for electron flow through the biofilm can be established (Liu and Bond 2012). The existence of such a gradient could be confirmed by Babauta et al. (2012b) in biofilms of approx. 200 µm thickness. The gradient increases as the biofilm thickens

(Babauta et al. 2012b). A redox gradient actively created by biofilm cells is another indicator for metabolic activity and contribution to current of all biofilm layers. Therefore, the linear correlation seen here between biofilm thickness and current density appears reasonable. However, this correlation is only valid as long as the substrate diffusion into the biofilm is not limited. Renslow et al. (2013) could show a linear correlation for a biofilm thickness of up to 170 µm in *G. sulfurreduens* pure cultures fed with 20 mM of acetate. Above this thickness, acetate concentration within the biofilm became limiting, which resulted in a slightly impaired current density. Yet, Renslow et al. (2013) could show that even though cells in close proximity to the anode were substrate depleted, they still served as a conductive network to transport electrons from the upper biofilm layers to the anode.

3.1.4 Electrochemical characterisation by cyclic voltammetry

Cyclic voltammetry (CV) was performed under turnover and non-turnover conditions to determine the formal redox potentials of c-type cytochromes involved in the electron transfer. Exemplary, CVs of a *G. sulfurreducens* pure culture recorded under catalytic conditions (turnover) and under acetate depletion (non-turnover) can be seen in **Figure 3.10**. Formal redox potentials of c-type cytochromes were extracted as described in chapter 2.4. For easier comparison with following experiments (see **Figure 3.12** below), the peaks were assigned letters from A to O.

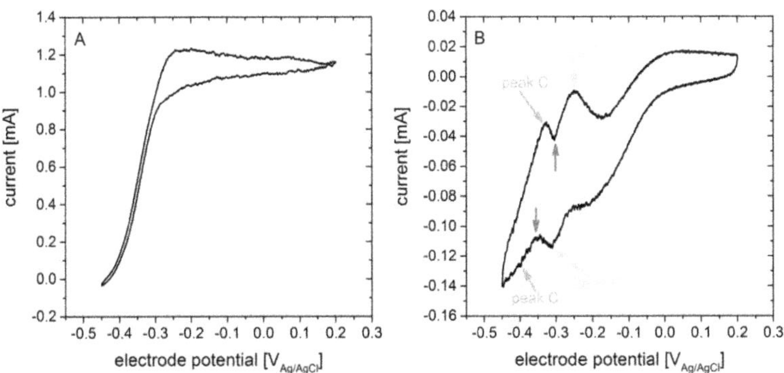

Figure 3.10: Exemplary (A) turnover and (B) non-turnover CVs of *G. sulfurreducens* recorded during cultivation cycle 2 at 0.25 mV/s in the range of -0.45 and 0.2 $V_{Ag/AgCl}$. Signals of peak C (orange) and D (yellow) are indicated in (B). Blue arrows indicate characteristic incision typically found in *Geobacter* spp. non-turnover CVs.

The signal for three peaks located at -0.363 ± 0.021 (peak B), -0.327 ± 0.025 (peak C) and -0.264 ± 0.021 $V_{Ag/AgCl}$ (peak D) was most intense and was present in all replicates. The high intensity indicates that these peaks belong to the c-type cytochromes mainly responsible for the electron transfer. Their abundance in the biofilm is probably very high, resulting in a strong signal. Consistent with this, Schmidt et al. (2017) found two main peaks in non-turnover CVs of biofilms consisting mainly of *G. anodireducens* located at -0.358 and -0.298 $V_{Ag/AgCl}$, respectively, which correspond well to peaks C and D. Occasional peaks were found at potentials of -0.178 ± 0.007 (F), -0.119 ± 0.007 (G) and 0.024 ± 0.025 $V_{Ag/AgCl}$ (I) (not indicated in **Figure 3.10**), but they did not appear reproducibly in all replicates. As signals from these peaks were very low, they may represent cytochromes that only play a minor role in current production. A peak at -0.08 $V_{Ag/AgCl}$ has been reported by Zhu et al. (2012) and could refer to the same c-type cytochrome as seen here at -0.119 $V_{Ag/AgCl}$. All of these peaks were detected under turnover conditions, while only peaks C, D and F were seen under non-turnover conditions. This is unusual as all cytochromes responsible for current production detected under turnover conditions should be expected to be seen under non-turnover conditions as well.

Examples of CVs from the defined mixed culture can be seen in **Figure 3.11**. In contrast to the *G. sulfurreducens* pure culture, only 2 peaks could be detected under turnover conditions. These were at -0.331 ± 0.014 and -0.266 ± 0.018 $V_{Ag/AgCl}$, which corresponds to peaks C and D. These peaks were also seen most abundant in *G. sulfurreducens* pure cultures, suggesting that *G. sulfurreducens* is responsible for the main current production in the defined mixed culture. Non-turnover CVs revealed 11 additional peaks ranging from -0.421 ± 0.014 to 0.3760 ± 0.0015 $V_{Ag/AgCl}$. Except for the two main peaks C and D, most of these were only found occasionally and not in all replicates. Usually, signals obtained in non-turnover CVs refer to cytochromes present in the biofilm but not responsible for the main current production.

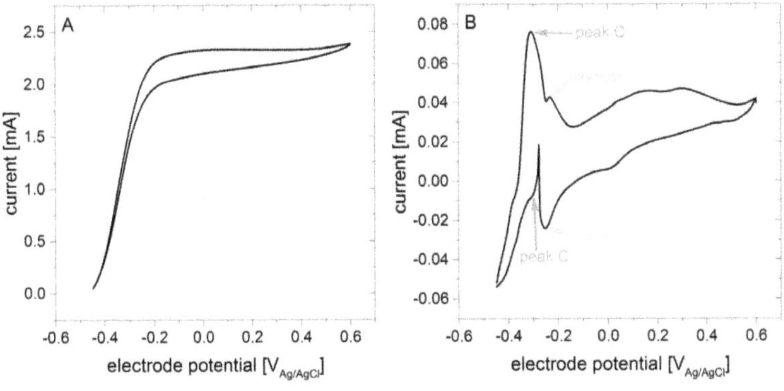

Figure 3.11: Exemplary (A) turnover and (B) non-turnover CVs of defined mixed culture of *G. sulfurreducens* and *S. oneidensis* recorded during cultivation cycle 1 at 0.25 mV/s in the range of -0.45 and 0.6 $V_{Ag/AgCl}$. Signals of peak C (orange) and D (yellow) are indicated in (B).

A comparison between the peaks found in the defined mixed culture and in the *G. sulfurreducens* pure culture can be seen in **Figure 3.12**. Three of the additional peaks were found at potentials above 0.2 $V_{Ag/AgCl}$, but *G. sulfurreducens* pure culture CVs were only recorded until this potential. Thus, it is unclear whether these peaks might be detectable in *G. sulfurreducens* pure cultures as well. Still, there were 4 more peaks detected in the mixed culture than in pure culture in the potential range between -0.45 and 0.2 $V_{Ag/AgCl}$. These additional peaks might be caused by the presence of *S. oneidensis*. Cytometry analysis revealed that the biofilm consisted mainly of *G. sulfurreducens*, thus, the additional peaks cannot be easily related to proteins produced by *S. oneidensis*. Rather, it is possible that they resulted from the expression of genes for additional c-type cytochromes by *G. sulfurreducens* that were caused by the presence of another species. It has been shown that co-cultivation of several species changes the transcription level of specific genes compared to pure cultures (Prokhorova et al. 2017). For example the expression of many genes involved in EET is upregulated in *G. sulfurreducens* as well as *S. oneidensis* under mixed culture conditions. Thus, the additional peaks found here may be a result of higher c-type cytochrome amounts from the dominant species *G. sulfurreducens*. The higher observed biofilm thickness may also be a cause for the additional peaks since in thicker biofilms redox potentials shift, as was discussed above in chapter 3.1.3.

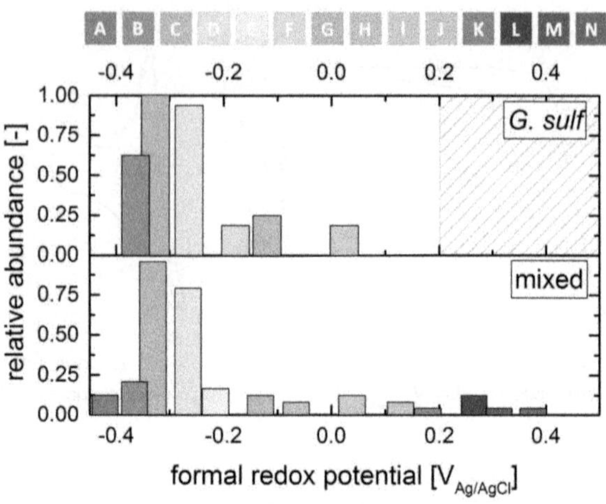

Figure 3.12: Relative abundance of formal potentials found in *G. sulfurreducens* pure culture (*G. sulf*) and defined mixed culture (mixed). Colours refer to peak assignment A to N as indicated. Grey area indicates the potential range not covered in pure culture CVs.

It is difficult to assign peaks detected in CV to individual molecules. Most cytochromes involved in electron transfer possess several heme groups that create a potential window of several hundred millivolts at which the cytochrome can operate (Santos et al. 2015). The potential windows of all cytochromes involved in electron transfer overlap, which allows for a smooth electron transfer from the periplasm to outer membrane cytochromes and to the extracellular electron acceptor (Santos et al. 2015). But the broad operation window of cytochromes makes it virtually impossible to assign individual cytochromes to peaks found in CV measurements. For example, OmcZ has a potential window of 405 mV and shows electrochemical activity between potentials of -0.619 and -0.214 $V_{Ag/AgCl}$, which means that each of the main peaks found here could stem from an OmcZ molecule (Santos et al. 2015). The same is true for other components of the electron transfer, for example OmcS, PpcA or MacA, whose formal redox potentials are -0.409 ± 0.16, -0.314 ± 0.142 and -0.385 ± 0.125 $V_{Ag/AgCl}$, respectively (Santos et al. 2015).

Liu and Bond (2012) found that there is a rate limiting step occurring in biofilms that have a thickness of over 10 µm located between potentials of -0.347 and -0.247 $V_{Ag/AgCl}$, which is where the main peaks C and D are located. In thin biofilms (< 10 µm), electron transfer occurs immediately and fast when the applied potential is changed (Liu and Bond 2012). In thick biofilms (> 20 µm), however, a portion of cytochromes will remain in the reduced state when potentials are shifted to more positive values (Liu and Bond 2012). This indicates a rate limiting step at these cytochromes located farther away from the electrode (Liu and Bond 2012). This rate limitation causes CV signals to split into several peaks (Liu and Bond 2012). Thus, the different redox potentials found here may simply refer to the same principal electron transfer mechanisms, but come from different depths of the biofilm created by altering electron transfer kinetics. One key feature that is different between thin and thick biofilms is the requirement of nanowires to maintain electron transfer (Reguera 2018). Cells deficient of producing pili are unable to grow beyond thicknesses of 10 µm (Reguera 2018). Above this thickness, pili become a necessary component for electron transfer from upper layers (Reguera 2018). Thus, the rate limiting step observed by Liu and Bond (2012) could be the transfer from c-type cytochromes to pili or vice versa.

Schmidt et al. (2017) could show that there is a reverse process taking place between the two main peaks C and D, creating the characteristic incision typically visible in *G. sulfurreducens*-based non-turnover CVs, indicated here by orange arrows in **Figure 3.10 B**. This reverse process during potential scans cannot be explained so far, but was connected by Schmidt et al. (2017) to a conformational change in c-type cytochromes that alters the properties of electron transfer. This change is observed in the same potential range as the rate limiting process described by Liu and Bond (2012), thus, there might be a connection of this conformational change to the charge transfer to cytochromes farther away from the electrode by means of nanowires.

One of the peaks detected in the defined mixed culture but not in pure culture is located at -0.421 ± 0.014 $V_{Ag/AgCl}$ (peak A). *S. oneidensis* produces flavins as electron shuttle (Marsili et al. 2008a), which it uses in the forms of riboflavin, FMN and FAD. The pure chemicals have formal redox potentials of -0.438, -0.443 and -0.443 $V_{Ag/AgCl}$, respectively (Hasford et al. 1997), which correspond well to the formal potential of peak A. In CVs from *S. oneidensis* pure cultures, Marsili et al. (2008a) found peaks centered at -0.42 $V_{Ag/AgCl}$ which they related to riboflavin. Thus, peak A found here might be caused by flavins produced by *S. oneidensis*. Considering the findings of Okamoto et al. (2014) that riboflavin increases current densities in *G. sulfurreducens* pure cultures by about 10 %, the production of flavins by *S. oneidensis* might contribute to the higher maximum current density observed here with the defined mixed culture.

Other components of the direct electron transfer pathway of *S. oneidensis* are the outer membrane cytochromes MtrC and OmcA. As these cytochromes have multiple heme centres similar to c-type cytochromes produced by *G. sulfurreducens*, their redox activity spans a great distance (Breuer et al. 2015). For example, MtrC shows activity between potentials of -0.7 and -0.1 $V_{Ag/AgCl}$ (Carmona-Martinez et al. 2011; Breuer et al. 2015). The formal potential of OmcA is found between -0.44 and -0.16 $V_{Ag/AgCl}$ (Marsili et al. 2008a; Meitl et al. 2009). These components might be responsible for peak E found in the defined mixed culture but not in *G. sulfurreducens* pure cultures at -0.215 ± 0.008 $V_{Ag/AgCl}$.

3.2 Influence of the anode potential on defined mixed culture

The optimal redox potential for cultivating electrochemically active bacteria in BESs has been investigated by several studies (Babauta et al. 2012a). Rather different results have been obtained by those studies as was summarised in chapter 1.6.2. To determine which of the observations from literature are valid here, the behaviour of the defined mixed culture was investigated at different anode potentials varying between -0.2 and 0.6 $V_{Ag/AgCl}$. Chronoamperograms and substrate concentrations behaved in similar ways as in the defined mixed culture operated at 0.2 $V_{Ag/AgCl}$. Current and substrate profiles varied greatly between replicates, possibly due to the spontaneity of the above discussed mutation that led to a consumption of lactate by *G. sulfurreducens*. Still, the same tendency of substrate consumption could be observed, whereby lactate was usually degraded prior to acetate in cultivation cycle 1 and acetate was degraded prior to lactate in cycle 2. The only exception was the experiment conducted at -0.2 $V_{Ag/AgCl}$, in which acetate was always degraded first. In cycles 2 and 3, the faster consumption of acetate was assumed to be possible due to the already established biofilm that allowed fast electron transfer to the anode. Thus, the prior consumption of acetate seen at -0.2 $V_{Ag/AgCl}$ in cycle 1 might be an indicator for easier electron transfer to the anode at this potential.

The maximum current densities achieved varied between experiments and are shown in **Figure 3.13**. Raw data of experiments can be found in the appendix (chapter 6, Figure A.1 to A.13). At 0.2 $V_{Ag/AgCl}$, the highest maximum current density could be achieved. Towards higher potentials, maximum current density decreased. This is in accordance with observations found by some authors for *S. oneidensis* pure cultures and undefined mixed cultures (Teravest and Angenent 2014; Zhu et al. 2014). In the work performed here, the lowest maximum current density was observed at 0.0 $V_{Ag/AgCl}$. Interestingly, when lowering the anode potential even further to -0.2 $V_{Ag/AgCl}$, a higher maximum current density was observed again. This result is not in accordance with literature. While the optimal potentials that have been found vary greatly, in all studies only one optimal potential range was found, instead of the two optima seen here (compare with **Table 1.1**).

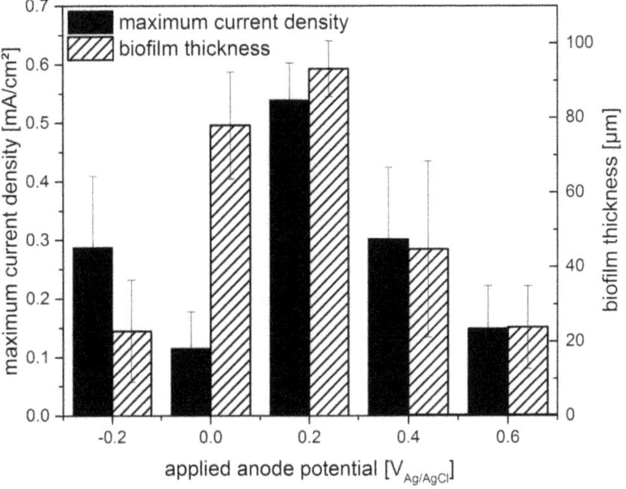

Figure 3.13: Maximum current density (filled bars) and biofilm thickness (diagonally ruled bars) of defined mixed culture experiments at various anode potentials.

Biofilms were analysed at the end of experiments and biofilm thickness was determined. The observed thickness at each potential can be seen in **Figure 3.13** together with the maximum current density. Biofilm thickness varies depending on the anode potential. The maximum biofilm thickness was observed at a potential of 0.2 $V_{Ag/AgCl}$. Increasing the potential led to thinner biofilms, concomitant with the lower maximum current densities seen. This is in accordance with the prior observation that maximum current density and biofilm thickness are correlated (see chapter 3.1.3). At 0.0 $V_{Ag/AgCl}$, however, biofilms grew to the same thickness as at a potential of 0.2 $V_{Ag/AgCl}$, while current production was reduced more than 4-fold. Very much the opposite is valid for -0.2 $V_{Ag/AgCl}$. While biofilms were only about as thick as at 0,6 $V_{Ag/AgCl}$, the maximum current density achieved at -0.2 $V_{Ag/AgCl}$ was almost twice as high.

This shows that the correlation between biofilm thickness and current density is not valid under all conditions and suggests that at low potentials the properties of biofilms forming is altered as they appear to deliver electrons more efficiently. This corresponds to the observed faster acetate consumption suggested above to be caused by better electron transfer. A possible explanation for the establishment of a biofilm better suited to transfer electrons might lie in the formal redox potentials of many insoluble iron-oxides, which belong to the natural electron acceptors for *G. sulfurreducens*. Their potentials range from -0.51 to -0.19 $V_{Ag/AgCl}$, thus the anode potential of -0.2 $V_{Ag/AgCl}$ used here offers the closest representation of the natural environment (Santos et al. 2015). Thus, in future experiments it is worthwile to investigate the effect of first conditioning cells at -0.2 $V_{Ag/AgCl}$ and later switching the potential to 0.2 $V_{Ag/AgCl}$ to achieve a thicker but on its base more productive biofilm. If successful, this strategy might improve current density and thus H_2 production rates even more.

The incorporation of *S. oneidensis* at different anode potentials was investigated by flow cytometry. There is evidence in literature that more *S. oneidensis* cells attach to electrodes at potentials higher than 0.2 $V_{Ag/AgCl}$ in undefined mixed cultures (Zhu et al. 2014), possibly because direct electron transfer mechanisms also function at these high potentials (Grobbler et al. 2018). This observation could not be confirmed here, however, as no *S. oneidensis* cells could be detected in the biofilms, regardless of the applied potential. Thus, even at potentials where biofilm formation and current production by *G. sulfurreducens* are hindered, *S. oneidensis* is outcompeted. Planktonically, amounts of *S. oneidensis* were highly diverse, ranging from 3 to 92 %, but there was no apparent connection of planktonic *S. oneidensis* to the applied potential.

CV measurements revealed most of the peaks found at 0.2 $V_{Ag/AgCl}$ which are collocated in **Figure 3.14**. At applied anode potentials between -0.2 and 0.4 $V_{Ag/AgCl}$, the two main peaks C and D are, again, predominant. The peaks E to M which were only occasionally found at 0.2 $V_{Ag/AgCl}$, are also seen, indicating that the expression of the corresponding genes for c-type cytochromes is not influenced by anode potential. For example peaks as high as peak M at 0.34 $V_{Ag/AgCl}$, were detected when -0.2 $V_{Ag/AgCl}$ was applied. This is an interesting result as cytochromes with high formal redox potentials have no means of contributing to current when the applied anode potential is very low. Still, the corresponding cytochromes were apparently produced by *G. sulfurreducens*. This suggests that *G. sulfurreducens* produces certain electron transfer proteins as default, regardless of the surrounding environment. Although this would mean a high metabolic burden for *G. sulfurreducens*, it could be an advantageous strategy to survive in environments where electron acceptors are scarce and span a wide range of potentials. With many different possible electron transfer pathways present at all times, there is virtually no adaptation time of *G. sulfurreducens* when one electron acceptor is depleted and another with a different redox potential is found. The possibility for fast adaptation is an advantage over other bacteria. A transcriptome or proteome analysis could confirm this hypothesis in future experiments.

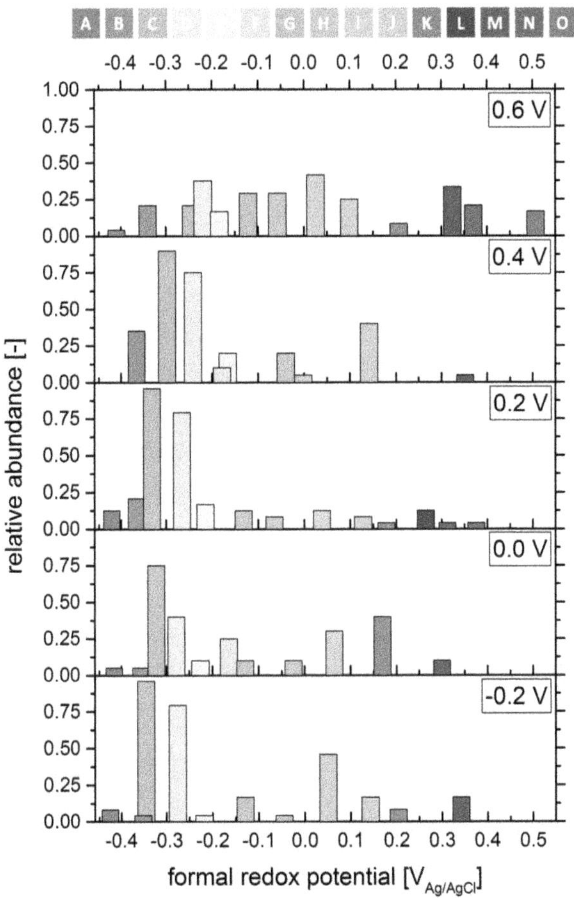

Figure 3.14: Relative abundance of formal potentials found in defined mixed culture conditioned at different anode potentials (vs. Ag/AgCl). Colours refer to peak assignment A to O as indicated. Numeric values can be found in Table A.1 in the appendix (chapter 6).

These findings are not in agreement with observations in literature made for undefined mixed cultures. Zhu et al. (2014) only found three peaks in CVs from mixed cultures conditioned at potentials between -0.45 and 0.31 $V_{Ag/AgCl}$. The formal potentials of these were at -0.38, -0.31 and -0.08 $V_{Ag/AgCl}$, respectively (Zhu et al. 2014). From the characteristic shape of *G. sulfurreducens* based CVs, the peaks located at -0.38 and -0.31 $V_{Ag/AgCl}$ found by these authors most likely correspond to peaks C and D found here, even though their respective potentials are lower than it was observed here. Considering this, the third peak found by Zhu et al. (2014) corresponds to peak H. However, the other 11 formal redox potential peaks reported here for similar applied anode potentials, were not found in that study.

Overall, CVs and formal potentials observed were very similar, both in the study conducted by Zhu et al. (2014) and in the present work, regardless of the applied potential in the range between -0.45 and 0.4 $V_{Ag/AgCl}$. But this was no longer valid when anodes were poised to 0.6 $V_{Ag/AgCl}$. At this potential, an additional peak, assigned peak O, appeared here which was located at 0.5066 ± 0.0024 $V_{Ag/AgCl}$. Such a high redox potential peak was also observed by Zhu et al. (2014) in a biofilm poised to 0.5 $V_{Ag/AgCl}$. A peak at similar potentials could also be found in pure cultures of *G. sulfurreducens* grown at 0.6 $V_{Ag/AgCl}$ (Busalmen et al. 2008; Zhu et al. 2012). Conditioning at 0.6 $V_{Ag/AgCl}$ further caused peaks present at lower anode potentials to weaken or even disappear (Busalmen et al. 2008; Zhu et al. 2012; Zhu et al. 2014). This agrees with observations made here. At the other applied anode potentials, peaks C and D are clearly the most dominant while the other peaks are only seen occasionally. Contrary to this, the relative amount of peaks found at 0.6 $V_{Ag/AgCl}$ is distributed evenly between the respective peaks.

The shape of turnover CVs reflects the same phenomenon (see **Figure 3.15**). Turnover CVs from biofilms conditioned at potentials below 0.6 $V_{Ag/AgCl}$ usually display a sigmoidal curve with a sharp increase in current between -0.35 and -0.25 $V_{Ag/AgCl}$ (see **Figure 3.15 A**). In between these potentials, peaks C and D are located which represent the c-type cytochromes responsible for the main current production and thus the usual electron transfer pathway of *G. sulfurreducens*. In turnover CVs from biofilms conditioned at 0.6 $V_{Ag/AgCl}$, however, the increase in current is not sharp, but rather gradual over the entire potential range (see **Figure 3.15 B**). The gradual increase indicates that many different cytochromes with different formal redox potentials contribute equally to current.

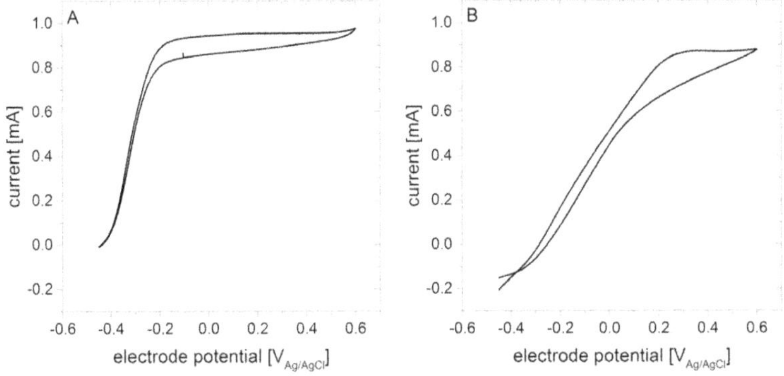

Figure 3.15: Exemplary turnover CVs from defined mixed culture biofilms conditioned at (A) -0.2 $V_{Ag/AgCl}$ and (B) 0.6 $V_{Ag/AgCl}$ recorded at 0.25 mV/s in the range of -0.45 and 0.6 $V_{Ag/AgCl}$.

As maximum current densities seen at 0.6 $V_{Ag/AgCl}$ were rather low, these transfer pathways appear to be less efficient than the usual pathway represented by peaks C and D. However, it is important to note that in turnover CVs recorded from biofilms conditioned at lower potentials the current did not decline when potentials of 0.6 $V_{Ag/AgCl}$ were achieved during the scan. This implies that the electron transfer pathway represented by peaks C and D is able to transfer electrons to an anode poised to 0.6 $V_{Ag/AgCl}$. Therefore, the reason for the low maximum current density seen at this potential does not lie in the inability of *G. sulfurreducens*' usual electron transfer system to cope with such high potentials. Rather, *G. sulfurreducens* is apparently unable to build this electron transfer network when constantly conditioned at 0.6 $V_{Ag/AgCl}$. This indicates that the potential has an influence on transcriptional and/or translational regulations, hindering the production of the efficient electron transfer network. Teravest and Angenent (2014) hypothesised that at redox potentials of 0.6 $V_{Ag/AgCl}$ cytochromes from *S. oneidensis* biofilms are denatured. If this is also the case for *G. sulfurreducens*, it would explain the lower maximum current density observed there. In further experiments, biofilms could be allowed to establish an efficient electron transfer pathway at lower anode potentials before being subjected to 0.6 $V_{Ag/AgCl}$ to distinguish between transcriptional or translational regulations and protein denaturation. Thus, it could be elucidated whether the same maximum current densities are achieved constantly (due to the established electron transport system) or whether current declines to values close to those observed here. The latter would be an indicator for the protein degradation proposed by Teravest and Angenent (2014).

3.3 Microaerobic growth of *Geobacter sulfurreducens*

3.3.1 Comparison of aerobic and anaerobic growth conditions

When the species *G. sulfurreducens* was originally found in 1994, the bacterium was considered to be a strict anaerobe as it did not grow in the presence of oxygen (Caccavo et al. 1994). Ten years later, however, it was discovered that oxygen can actually be reduced by *G. sulfurreducens* and even supports cell growth (Lin et al. 2004). The investigations of this first study were quite general. For example, oxygen concentrations were only measured in the gaseous head-space and not in the liquid phase so that the concentrations actually encountered by *G. sulfurreducens* were unknown (Lin et al. 2004). Therefore, *G. sulfurreducens* was cultivated here in a bioreactor system that allowed the controlled and continuous addition of varying concentrations of oxygen. An anaerobic control experiment was conducted for comparison in which 40 mM of fumarate served as electron acceptor (experiment abbreviation 40F0%). For microaerobic experiments, fumarate was replaced by oxygen. It was already shown by Lin et al. (2004) that initial cell growth with fumarate is necessary for *G. sulfurreducens* to be capable of oxygen reduction. Hence, microaerobic experiments were performed with an anaerobic adjustment period of 2 h, in which 4 mM of fumarate served as electron acceptor. Subsequently, the oxygen concentration in the gas inlet was changed to 1, 3 or 5 % (experiment abbreviation 4F1%, 4F3% and 4F5%, respectively). An additional, completely anaerobic cultivation with only 4 mM of fumarate was also conducted (experiment abbreviation 4F0%) to ensure that cell growth observed in microaerobic experiments was not due to the provided fumarate.

The resulting cell growth curves of all experiments are shown in **Figure 3.16**. Exponential growth at a specific growth rate of $0.109 \pm 0.015\ h^{-1}$ could be observed in the anaerobic control experiment 40F0% and a maximum cell density of $0.417 \pm 0.008\ g_{CDW}/L$ was obtained. Under fumarate limited conditions (4F0%), cell growth was strongly reduced and linear instead of exponential. When 1 % of oxygen was introduced as additional terminal electron acceptor (4F1%), a likewise linear cell growth was observed, but the slope of this growth was higher compared to 4F0%. As the remaining cultivation conditions were the same, oxygen is clearly the reason for the enhanced cell growth. Increasing the oxygen concentration to 3 % resulted in even faster linear cell growth (4F3%). This indicates that growth rates are directly dependent on the provided amount of oxygen. The maximum biomass concentration achieved in experiment 4F3% was $0.415 \pm 0.006\ g_{CDW}/L$ and was thus the same as in the control experiment 40F0%. This shows that oxygen allows the same cell growth with acetate as fumarate does. In experiment 4F5%, the oxygen concentration introduced was too high, so that cell growth was completely inhibited.

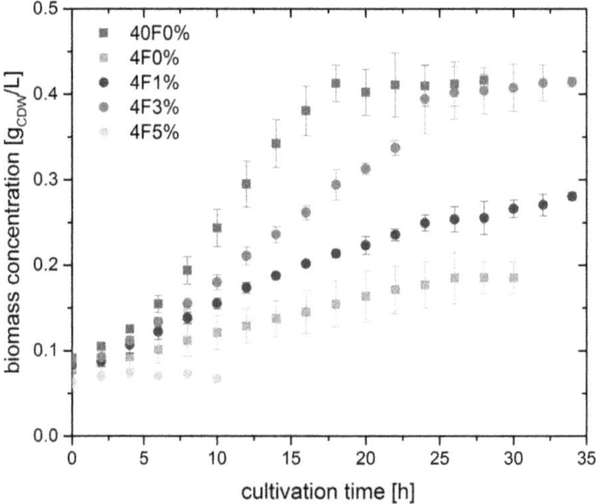

Figure 3.16: Course of biomass concentration of *G. sulfurreducens* during growth with 10 mM of acetate under anaerobic (squares) and microaerobic (circles) conditions at 30 °C. Error bars represent standard deviation from biological triplicates.

The concentration of dissolved oxygen (DO) was monitored online during cultivations and levels of acetate were determined via HPLC analysis. Both are exemplary shown for experiment 4F3% in **Figure 3.17**. Until 24 h after inoculation, the linear increase in biomass is associated with a linear decrease in acetate concentration. During this growth period, the DO remains at 0 mg/L except for a brief spike at 2 h, which marks the time point of the change in gas composition. This indicates that all introduced oxygen was reduced by the cells. Principally the same was observed in experiment 4F1%. Oxygen was provided continuously at the same rate and since all oxygen was reduced, it was limiting for cell growth. This limitation explains the linearity in biomass production and acetate consumption observed.

Another possibility for oxygen reduction is the reaction with a different chemical species. The only chemical species in the medium capable of oxygen capturing is cysteine, but the cysteine concentration of 2 mM cannot suitably explain the reduction of oxygen. Further, as DO sensors were equilibrated overnight at 21 % O_2 for calibration, cysteine should already be completely reduced at the beginning of cultivations. Therefore, only the cells remain as cause for oxygen reduction.

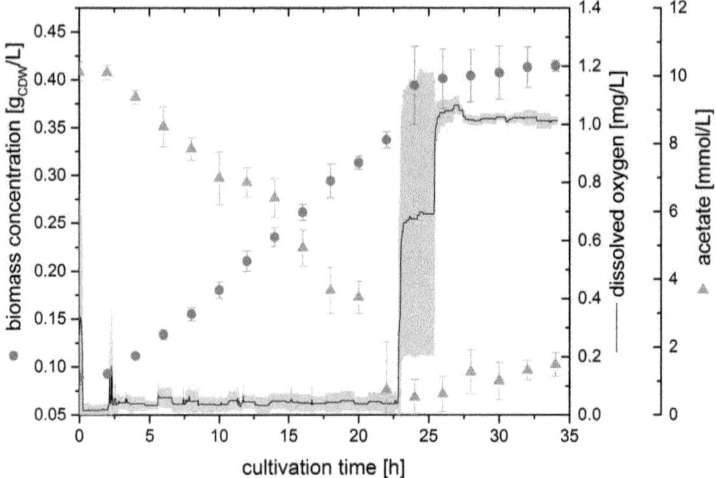

Figure 3.17: Biomass concentration (circle), acetate (triangle) and dissolved oxygen (line) concentration during growth of *G. sulfurreducens* with 10 mM of acetate and 3 % of oxygen in the gas inlet (experiment 4F3%). Error bars and grey area represent standard deviation from biological triplicates.

When acetate consumption and cell growth subsided, the DO sharply increased to levels around 1 mg/L. The increase did not occur at precisely the same time in each bioreactor, creating the large standard deviation visible in **Figure 3.17**. When acetate is exhausted, a carbon and energy source to support cell growth is lacking. No more electrons are present for oxygen reduction, thus, the introduced oxygen stays dissolved in the medium. The maximum DO concentration c_{O2}^* for the medium was determined to be 0.31 mg/L per % of oxygen in the gas inlet. Thus, when 3 % of oxygen are used, c_{O2}^* should be 0.93 mg/L. The DO concentration of 1 mg/L present after acetate depletion agrees well with this value.

When 5 % of oxygen was provided, there was no cell growth at all nor was there any acetate consumption. The DO concentration increased to 1.4 mg/L, which is in the range of the expected maximum level of 1.6 mg/L. Therefore, it can be assumed that no oxygen reduction occurred in this experiment. This implies that the oxygen concentration provided is too high to be manageable by *G. sulfurreducens*. Interestingly, cells are not capable of consuming oxygen only partially or of growing on the provided amount of fumarate. There appears to be a maximum amount of oxygen reducible by the cells. If this maximum value is surpassed, cell growth is completely inhibited. Microaerobic growth in *G. sulfurreducens* can therefore be termed as an all-or-nothing condition.

G. sulfurreducens was reported previously by Lin et al. (2004) to reduce oxygen. They described consumption of oxygen from the head space of cultivation flasks along with cell growth and acetate consumption. Similar to the results presented here, they observed a dependency of the capability of oxygen reduction on the oxygen concentration provided. In their case, oxygen reduction coupled to cell growth was only achieved with 5 % or 10 % in the gaseous phase of the headspace of culture flasks. However, DO was not monitored in the medium (Lin et al. 2004), hence no quantitative comparison of the results is possible here.

5 % of oxygen provided by the gas inlet were found to be growth limiting under the conditions tested here. But in order to determine whether this is an absolute limit or whether the ability of oxygen reduction is rather dependent on the biomass concentration, an additional experiment was performed. Hereby, the amount of introduced oxygen was adapted in a stepwise manner depending on the concentration of cells present (experiment abbreviation 4FSI). To achieve this, the maximum specific oxygen uptake rate (*sOUR*) was calculated with equation (3.1), using the k_La-value and maximum DO concentration c_{O2}^* determined for the system and the biomass concentration X present at 2 h of cultivation from experiment 4F3%. This timepoint and experiment were chosen as the amount of introduced and reduced oxygen and thus the *sOUR* were highest here.

$$maximum\ sOUR = \frac{k_La \cdot c_{O2}^*}{X} = \frac{9.4 \pm 1.1\ h^{-1} \cdot 0.935\ \frac{mg_{O2}}{L}}{0.0928 \pm 0.0024\ \frac{g_{CDW}}{L}} = 95\ \pm 11\ \frac{mg_{O2}}{h \cdot g_{CDW}} \tag{3.1}$$

The amount of introduced oxygen was adjusted after each sampling and OD_{600} determination so that the *sOUR* remained at this level throughout experiment 4FSI. Biomass and acetate as well as DO concentration obtained are shown in **Figure 3.18**. Like in the forgoing experiments, the DO concentration stayed at 0 mg/L as long as acetate was present in the medium. Due to the constant increase in the amount of oxygen provided, exponential cell growth could be observed at a specific growth rate of 0.124 ± 0.007 h^{-1}. This is in the same range as determined for the control experiment 40F0%, which shows that *G. sulfurreducens* can use oxygen as efficiently as fumarate. By the end of the cultivation, a gas inlet concentration of 12.5 % of oxygen could be achieved, which is a 2.5-fold increase compared to the beforehand observed growth limiting oxygen concentration of 5 %. This demonstrates that the biomass concentration is determining the amount of oxygen reducable by *G. sulfurreducens*. This result is novel to the work of Lin et al. (2004), who just demonstrated continuous cell growth with 5 % of oxygen repeatedly provided in the headspace of cultivations. The maximum biomass concentration achieved here was 0.473 ± 0.018 g_{CDW}/L after 20 h of cultivation. This value is 13 % higher than the maximum biomass concentration achieved in experiment 40F0%, and suggests that oxygen allows for a more efficient substrate conversion into biomass.

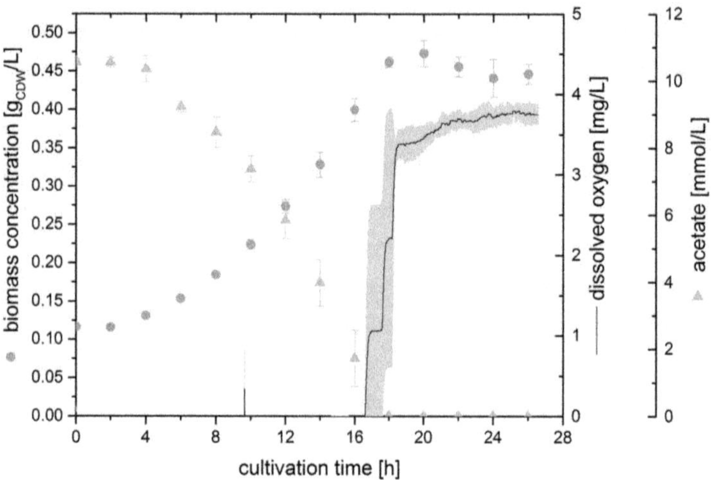

Figure 3.18: Biomass concentration (circle), acetate (triangle) and dissolved oxygen (line) concentration during growth of *G. sulfurreducens* with 10 mM of acetate and stepwisely increasing amounts of oxygen in the gas inlet (experiment 4FSI). Error bars and grey area represent standard deviation from biological triplicates.

3.3.2 Metabolisation of selected organic acids

Acetate was metabolised in all experiments for biomass formation and energy production. In *G. sulfurreducens*, ATP is generated solely by electron transport phosphorylation (Galushko and Schink 2000; Lovley et al. 2004). Hereby, NADH formed in the TCA cycle is oxidised by the type I NADH dehydrogenase which simultaneously transports protons into the periplasm, creating a proton gradient for ATP synthesis (Mahadevan et al. 2006). Electrons derived here are fed into the membrane bound menaquinon pool of *G. sulfurreducens* (Caccavo et al. 1994; Galushko and Schink 2000; Butler et al. 2006). From menaquinol, electrons are transferred to the terminal electron acceptor, in the case of the experiments performed here either fumarate or oxygen. The energy potentially available for ATP production is determined by the difference in redox potential $\Delta E°'$ between electron donor and acceptor and is greater for oxygen ($E°'$ = 815 mV, Wood 1988) than for the redox couple fumarate/succinate ($E°'$ = 30 mV, Butler et al. 2006). Thus, oxygen might allow *G. sulfurreducens* to use acetate more effectively. An improved conversion of acetate into biomass would result in a higher biomass to acetate yield $Y_{X/S}$, which was therefore determined from the linear slope of biomass and acetate concentration during cell growth plotted against each other. In the comparative cultivation 40F0%, $Y_{X/S}$ was determined to be 0.552 ± 0.014 g/g and similar values were obtained for most of the other cultivations. The only exception was experiment 4F1% which resulted in a marginally larger $Y_{X/S}$ of 0.63 ± 0.06 g/g. This suggests that oxygen could actually cause a more efficient acetate utilisation, but only as long as the *sOUR* is well below the maximum. Oxygen eventually becomes growth inhibiting, so it does have a negative effect on *G. sulfurreducens*. Thus, the primary goal of *G. sulfurreducens* when confronted with oxygen will be to remove all oxygen, while the potential energetic benefit is only secondary. This could explain the return of $Y_{X/S}$ to 0.55 g/g in experiments 4F3% and 4FSI. The results are comparable to observations from Lin et al. (2004) who also saw equal acetate to biomass yields with either oxygen or fumarate as terminal electron acceptor.

Next to acetate, the concentration profiles of fumarate, pyruvate, succinate and 2-oxoglutarate were also determined. 2-oxoglutarate could not be detected in substantial quantities and will hence not be discussed further. From fumarate concentrations, the yield of fumarate reduced per acetate $Y_{Fum/Ac}$ was calculated for the anaerobic control experiment 40F0% to be 2.53 ± 0.04 mol/mol. If acetate is reduced completely to CO_2, 8 electrons need to be transferred to fumarate. As fumarate reduction to succinate consumes only 2 electrons, 4 fumarate molecules are needed to oxidise 1 molecule of acetate. Thus, the stoichiometry of fumarate reduction to acetate oxidation is 4:1. Not all acetate is oxidised for energy production, however, which is why the actual stoichiometry is lower as seen with 2.53:1 here. This indicates that 63 % of acetate (calculated by 2.53/4) were used for energy production, while 37 % were used for biomass formation. A slightly different ratio was shown by Esteve-Núñez et al. (2005), who determined that 75 % are used for energy production, while 25 % of acetate are used for biomass formation when fumarate is the electron acceptor.

A similar yield can be calculated in the microaerobic cultivations 4F1%, 4F3% and 4FSI for oxygen reduction ($Y_{O2/Ac}$). As one molecule of oxygen requires 4 electrons to be reduced, the stoichiometry for complete acetate oxidation with oxygen is 2:1. Seeing as the same $Y_{X/S}$ was determined for aerobic and anaerobic experiments, $Y_{O2/Ac}$ should be expected to be in the same range as $Y_{Fum/Ac}$. However, the determined $Y_{O2/Ac}$ for cultivations 4F1%, 4F3% and 4FSI is at 1.64 ± 0.13, 1.71 ± 0.13 and 1.92 ± 0.09 mol/mol, respectively. This would suggest that 82 to 96 % of acetate was used for energy production and less acetate (only 4 % to 18 %) was used for biomass formation compared to experiment 40F0%. The consumption of succinate in aerobic experiments (see below) provides an explanation for this inconsistency. Succinate metabolised via the TCA cycle results in 2 additional NADH equivalents, which can be used to reduced oxygen instead of electrons derived from acetate. Thus, $Y_{O2/Ac}$ could be erroneously increased.

The concentrations of fumarate and succinate are shown in **Figure 3.19** for cultivations 4F0%, 4F1%, 4F3% and 4FSI. In the beginning of cultivations, there is a consumption of fumarate and a concurrent generation of succinate. This trend is ongoing in experiments 4F0% and 4F1% until fumarate is completely depleted. The same principle observation was made in experiment 40F0%. Fumarate serves as terminal electron acceptor under anaerobic conditions and is reduced to succinate with a bifunctional fumarate reductase / succinate dehydrogenase at a stoichiometry $Y_{Suc/Fum}$ of 1:1 (Butler et al. 2006). The TCA cycle is working as an open loop under these conditions (see black arrows in **Figure 3.20**) (Galushko and Schink 2000; Butler et al. 2006; Mahadevan et al. 2006). In experiment 40F0%, the stoichiometry observed was 0.972 ± 0.025 mol/mol, which corresponds well to the expected value. But when electron acceptor limiting conditions are created in experiment 4F0%, $Y_{Suc/Fum}$ increases to 1.30 ± 0.05 mol/mol, indicating that more succinate is formed than should be possible from fumarate. This suggest other substrates to be used as precursor for fumarate to generate more electron acceptor. For instance, decreasing pyruvate concentrations (see below) suggest an enhanced flow of carbon via anaplerotic reactions that in the end form fumarate (see orange arrows in **Figure 3.20**). The carbon for these reactions might be provided by cysteine present in the medium as *G. sulfurreducens* can metabolise cysteine to pyruvate (Methé et al. 2003).

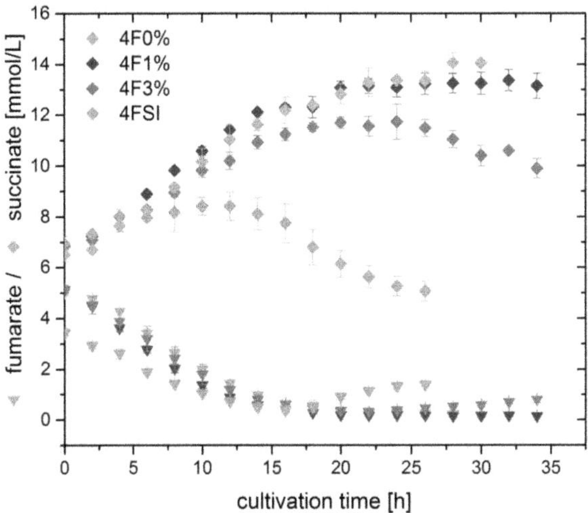

Figure 3.19: Concentration of fumarate (triangles) and succinate (diamonds) during growth of *G. sulfurreducens* with 10 mM of acetate, 4 mM of fumarate and varying concentrations of oxygen. Error bars represent standard deviation from biological triplicates.

In contrast, in cultivations 4F3% and 4FSI succinate concentration is declining again after 24 h and 10 h, respectively. The *sOUR* was relatively high in experiment 4F3% and was at its maximum in experiment 4FSI. Thus, succinate is being consumed when oxygen is provided and this consumption increases with the *sOUR* (see **Figure 3.19**). With oxygen as terminal electron acceptor, the bifunctional fumarate reductase / succinate dehydrogenase works in the opposite direction so that the TCA cycle can function as a closed loop (see blue arrows in **Figure 3.20**). Oxidation of succinate to oxaloacetate or pyruvate produces NADH and menaquinol which can be used to reduce oxygen (Butler et al. 2006). It was seen from DO concentrations during experiments that oxygen must be reduced completely in order for cells to grow. Failing to reduce oxygen results in a complete inhibition of cell growth, as seen from experiment 4F5%. This suggests *G. sulfurreducens* needs to dispose of any oxygen in its vicinity. Therefore, the carbon flow could be shifted from succinate down the TCA cycle to create more reducing equivalents for oxygen reduction.

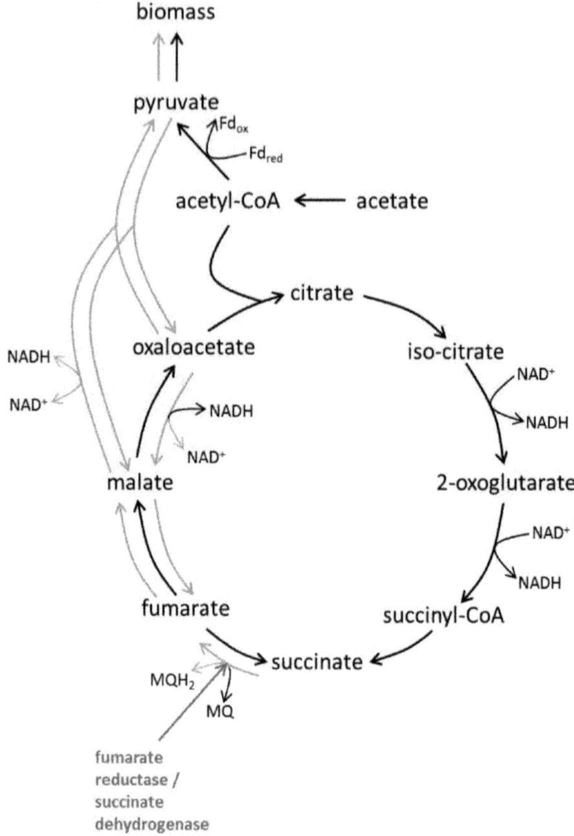

Figure 3.20: Flow of carbon around the tricarboxylic acid (TCA) cycle under fumarate respiring conditions (black arrows) as derived from Butler et al. (2006) and Segura et al. (2008) and proposed alterations in carbon flow during electron acceptor limiting (orange arrows) and microaerobic growth supporting (blue arrows) conditions. Fd = ferrodoxin, MQ = menaquinone, MQH_2 = menaquinol. Red annotations refer to enzymes responsible for the indicated reactions.

The concentrations of pyruvate are shown in **Figure 3.21**. Pyruvate was not supplemented to the medium, but was present in the inoculum. Three different behaviours are seen for pyruvate. In cultivations 4F0% and 4F1% pyruvate is degraded, corresponding to the above discussed need for anaplerotic reactions to provide more fumarate. Contrary, in cultivation 40F0%, pyruvate levels rise to 3.3 mM. With fumarate present in excess in this experiment, there is no need for enhanced anaplerotic reactions.

In experiment 4FSI, pyruvate is accumulating in the medium to a concentration of 0.73 mM. Similarly, pyruvate accumulates in experiment 4F3% in the initial 10 h. This indicates that accumulation of pyruvate is occuring when the *sOUR* is high and agrees with the consumption of succinate in experiments 4FSI and 4F3% discussed above. A direction of succinate carbon to pyruvate would function as additional carbon source for cell growth. In experiment 4FSI, 3 mM of succinate are oxidised, but only 0.5 mM of pyruvate are formed. If a 1:1 stoichiometry for this reaction and no additional carbon flow to oxaloacetate are assumed, this leaves a potential of 2.5 mM pyruvate to promote biomass production (indicated by blue arrows in **Figure 3.20**). This explains the 13 % higher biomass concentration found in experiment 4FSI. But to account for all carbon, a more detailed analysis of metabolic fluxes, for example by determining intracellular metabolite levels, as well as investigating the elementary composition of *G. sulfurreducens* are necessary.

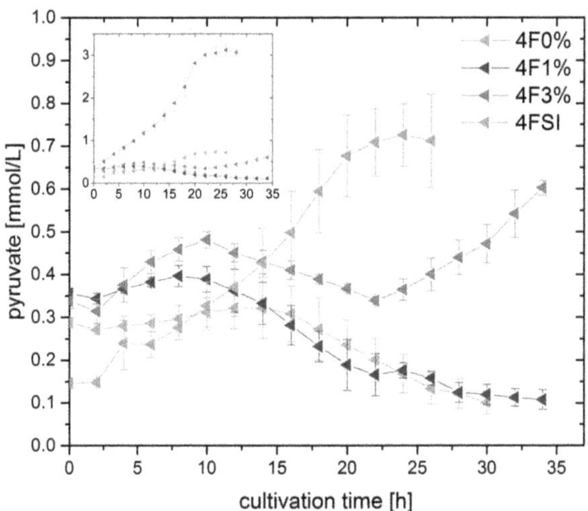

Figure 3.21: Concentration of pyruvate during growth of *G. sulfurreducens* with 10 mM of acetate, 4 mM of fumarate and varying concentrations of oxygen. Inset graph additionally shows pyruvate concentration during anaerobic growth with 40 mM of fumarate (experiment 40F0%). Error bars represent standard deviation from biological triplicates.

3.3.3 Transcriptome analysis under anaerobic and microaerobic conditions

Genome analysis of *G. sulfurreducens* showed several proteins possibly involved in oxygen reduction (Methé et al. 2003). The expression of many of the corresponding genes were found to be regulated by RpoS, a regulon required by *G. sulfurreducens* for cell growth with oxygen (Nunez et al. 2006). Located within RpoS are, for example, genes for the cytochrome c oxidase, a cytochrome bd menaquinol oxidase and a putative rubredoxin:oxygen oxidoreductase (Nunez et al. 2006). The cytochrome c oxidase is part of the usual oxidative phosphorylation pathway present in aerobic bacteria and was therefore expected to be the enzyme enabling oxygen reduction (Methé et al. 2003; Lin et al. 2004). This hypothesis was supported by further work of Lin (unpublished results), who according to Núñez et al. (2006) showed that *G. sulfurreducens* loses the ability to grow with oxygen if the cytochrome c oxidase is deleted. But in another representative of the *Geobacter* family, *G. uraniireducens*, the gene for a cytochrome bd menaquinol oxidase was also expressed at higher levels following exposure to oxygen (Mouser et al. 2009). Other studies have investigated the consequences of oxygen exposure on genes of *G. sulfurreducens* associated with oxidative stress (DiDonato et al. 2006; Tremblay and Lovley 2012). But the mechanism by which *G. sulfurreducens* reduces oxygen has not been studied in detail. Therefore, a transcriptome analysis was performed with RNA derived from cultivations 4F1%, 4F3% and 4F5%, whereby RNA from cultivation 40F0% served as comparison to elucidate possible oxygen reduction mechanisms.

Experiment 4F3% revealed 19 genes that were expressed at higher levels compared to 40F0%, while the expression of only one gene was found to be downregulated. In 4F1% and 4F5%, the expression of a total of 85 and 558 genes was upregulated while 33 and 483 genes were expressed at lower levels, respectively. A selection of the genes differentially expressed is shown in **Table 3.1**. Interestingly, the cytochrome c oxidase, which was mentioned above as the most likely enzyme responsible for oxygen reduction, was not expressed at higher levels under any of the conditions tested. Rather, the expression of the gene GSU1641 was found upregulated up to 2.3-fold in cultivations 4F3% and 4F1%. This gene encodes the catalytic subunit II of the cytochrome bd menaquinol oxidase which is capable of reducing oxygen to water with the concurrent oxidation of menaquinol to menaquinon. It represents the same oxidase that was also found to be part of RpoS by Núñez et al. (2006), and was found upregulated in *G. uraniireducens* after oxygen exposure by Mouser et al. (2009). This implies that the menaquinol oxidase encoded by GSU1641 is the more important enzyme for oxygen reduction in *G. sulfurreducens* compared to the cytochrome c oxidase. Likewise, the gene for the putative rubredoxin:oxygen oxidoreductase mentioned above was not found differentially expressed. It cannot be excluded that these other potential oxygen reducing enzymes also play a role in oxygen reduction, but for a complete consumption *G. sulfurreducens* only expresses the gene for the menaquinol oxidase at higher levels. A possible explanation for this could be that this enzyme is a high-oxygen affinity terminal oxidase (Methé et al. 2003; Nunez et al. 2006), which would ensure efficient and fast oxygen removal.

It was investigated above, whether the greater potential difference $\Delta E^{\circ\prime}$ between electron donor and acceptor during oxygen respiration compared to fumarate would result in more efficient biomass formation (see chapter 3.3.2). However, this was not observed. As demonstrated in chapter 1.6.3, the energy gain from an electron donor will depend on the ability of the cell to turn the energy residing in $\Delta E^{\circ\prime}$ into a proton gradient. The cytochrome c oxidase, which represents complex IV of the oxidative phosphorylation usually used by aerobic bacteria, is capable of translocating 3 protons per electron consumed (Kabashima et al. 2009). But the cytochrome bd menaquinol oxidase only generates proton motive force equivalent to 1 proton translocation per electron (Kabashima et al. 2009). It is thus no more efficient at generating proton motive force than is the fumarate reductase (Mahadevan et al. 2006). This explains why biomass formation from acetate is not more efficient with oxygen as terminal electron acceptor.

Evidence that cytochrome bd menaquinol oxidase is the responsible enzyme for oxygen reduction can even be derived from electrochemical experiments. In the work of Prokhorova et al. (2017) discussed above in the context of electrochemical pure vs. mixed cultures, the genes GSU1641 and GSU1640 corresponding to the two subunits of cytochrome bd menaquinol oxidase were observed to be expressed at lower levels under mixed culture conditions. This indicates that in the presence of other organisms, *G. sulfurreducens* relies on facultative anaerobes to reduce oxygen. It only makes sense to renounce the transcription of these genes if they actually serve a purpose, in this case oxygen reduction, which is no longer necessary under mixed culture conditions.

*Table 3.1: Selected genes expressed differentially under microaerobic conditions. Differentially expressed genes with p-value > 0.05 and > 0.1 are marked * and **, respectively. 1%, 3% and 5% refer to experiments 4F1%, 4F3% and 4F5%, respectively, + = higher expression, - = lower expression, e = not differentially expressed, FC = fold change.*

Locus tag	Protein name		+/- 1%	+/- 3%	+/- 5%	log_2 FC 1%	log_2 FC 3%	log_2 FC 5%
GSU1641	Catalytic subunit II	Cytochrome bd menaquinol oxidase	+	+	e	1.18*	1.12	0.63
GSU1640	Subunit I		e	e	-	0.28	0.44	-1.07
GSU0219	Cytochrome c oxidase		e	e	e	0.6	0.05	0.61
GSU0220			e	e	e	0.39	0.13	-0.42
GSU0221			e	e	e	0.61	0.12	-0.19
GSU0222			e	e	e	0.34	0.07	-0.5
GSU3294	Rubredoxin:oxygen oxidoreductase		e	e	e	0.42	0.68	0.07
GSU2029	PilP	type IV pili	(+)	e	-	0.88	0.58	-1.33
GSU2030	PilO		(+)	e	-	0.99	0.61	-1.11
GSU2031	PilN		(+)	e	(-)	0.87	0.52	-0.84
GSU2032	PilM		+	e	e	1.21**	0.61	-0.66
GSU2034	PilX-2		+	e	e	1.53	0.69	-0.5
GSU2035	PilW-2		+	e	e	1.59	0.72	-0.37
GSU2036	PilV-2		+	e	-	1.93	0.79	-1.59
GSU2037	FimU		+	e	-	1.76	0.71	-1.22
GSU2038	PilY1-2		+	(+)	e	1.73	0.88	-0.38
GSU2039	PilL		+	e	e	1.41	0.7	-0.32
GSU0036	Poly-γ-glutamate capsule biosynthesis protein		e	e	+	0.03	0.12	2.2
GSU1835	Glutamine synthetase		e	e	+	-0.27	-0.11	1.14
GSU1239	Glutamate synthase		e	e	+	-0.57	0.46	1.83
GSU1467	2-oxoglutarate ferredoxin oxidoreductase		e	e	+	-0.62	-0.08	-5.71
GSU1468			e	e	+	-0.38	-0.01	-4.69
GSU1469			e	e	+	-0.56	-0.1	-4.82
GSU1470			e	e	+	-0.56	-0.06	-4.79
GSU3356	Diguanylate cyclase		e	e	+	-0.04	-0.03	2.02
GSU2828			e	e	+	0.24	0.27	1.14
GSU0078	PilZ domain containing protein		e	e	+	0.08	0	1.31
GSU2089	Rod shape determining protein		e	e	+	0.28	-0.02	-1.54
GSU2193	Ferritin-like domain-containing protein		e	e	+	0.59	0.01	1.62
GSU2967	Ferritin-like domain-containing protein		+	e	+	1.67	0.30	3.17
GSU0720	Desulfoferrodoxin		e	e	+	0.32	0.16	1.53

Other parts necessary for energy production, for example genes for the ATP synthase or the NADH dehydrogenase were not differentially expressed in cultivations 4F3% and 4F1%. There were also no alterations seen with the genes for most enzymes of the TCA cycle and the glycolysis/gluconeogenesis pathway. This indicates that cell growth and metabolism work principally in the same manner as in an anaerobic environment when oxygen serves as terminal electron acceptor.

In contrast, many genes of the oxidative phosphorylation, the TCA cycle and the glycolysis/gluconeogenesis were expressed at lower levels in experiment 4F5%. This is consistent with the inhibition of cell growth observed. Instead, the expression of genes related to oxidative stress was seen upregulated up to 9-fold here. These genes include for example the ferritin-like domain containing proteins GSU2193 and GSU2967. Also, GSU0720 was expressed at higher levels (2.9-fold), which encodes for desulfoferrodoxin, an enzyme responsible for the reduction of superoxide (O_2^-) to H_2O_2 and found in multiple anaerobic bacteria for detoxification (Jenney Jr. 1999; Lombard et al. 2000). This indicates that *G. sulfurreducens* focuses on protecting itself against radical oxygen species when oxygen concentrations are too high to allow for a complete reduction.

Further results from the transcriptome analysis indicate that *G. sulfurreducens* can control its reaction to oxygen depending on the *sOUR* present and applies different strategies for survival. In cultivation 4F1% the expression of 11 type IV pilus genes (GSU2029-GSU2039) was upregulated up to 3.8-fold. The functions of type IV pili range from cell communication, adhesion and biofilm formation to cell motility (Tran et al. 2008; Daum and Gold 2018). Motility is induced by the protrusion of pilus fibers and the attachment of their tips on surfaces, followed by the retraction of the pilus, which pulls the cell body forward (Burrows 2012). In *G. sulfurreducens*, pili induced motility has not been described. Rather, pili have the function of constituting conductive nanowires for extracellular electron transport to insoluble electron acceptors like iron oxide particles (Reguera 2018). Fumarate and oxygen are both limiting for cell growth in experiment 4F1%, thus, a higher expression of genes for pili could represent an attempt to find other, insoluble electron acceptors. However, most of the genes associated with nanowires, especially that for PilA, the major pilin responsible for conductivity, are found elsewhere in the genome (GSU1491-GSU1497). All of these genes are not expressed differently, however.

The genes GSU1491 to GSU1497 have been studied by others because of their relation to the conductive nanowires (Reguera et al. 2005; Richter et al. 2012; Steidl et al. 2016). But there are no studies that dealt specifically with the genes GSU2029 to GSU2039 in *G. sulfurreducens*, so only speculations about their exact functions are possible here. GSU2029 to GSU2039 include *pil*M, *pil*N, *pil*O and *pil*P, which encode for proteins hypothesised to be necessary components of the assembly for pilus protrusion and retraction in *G. sulfurreducens* (Reguera

2018) because they serve this function in other bacteria (Burrows 2012; Daum and Gold 2018). Also included are genes that encode for minor pilins. The role of minor pilins is not yet clarified entirely, but in *Pseudomonas aeruginosa* they form complexes from which a new pilus protrusion can be initiated, whereby the minor pilins form the tip of the pilus fibre (Burrows 2012). If the concentration of minor pilins in the membrane is raised, it results in more and shorter pili on the cell surface (Burrows 2012). More pili might allow for attachment of *G. sulfurreducens* on a surface and since pili serve the function of motility in other bacteria, their purpose could be to escape the microaerobic environment. Another indicator for this can be found in the study of Mouser et al. (2009), who saw motility related genes upregulated in *G. uraniireducens* after exposure to oxygen. Motility caused by protrusion and retraction of pili is energy intensive, which might be why *G. sulfurreducens* only applies this strategy when the *sOUR* is low enough that finding an anaerobic zone is actually likely. Consequently, the higher expression of genes GSU2029 to GSU2039 is no longer seen in cultivations 4F3% or 4F5%. Whether type IV pili induced motility is indeed occuring in *G. sulfurreducens* has to be investigated in further studies.

Contrary to a strategy of fleeing, genes expressed at higher levels in cultivation 4F5% are encoding for proteins involved in biofilm formation and encapsulation. For instance, the expression of the gene for a poly-γ-glutamate capsule biosynthesis protein (GSU0036) was upregulated 4.6-fold. Production of a poly-γ-glutamate capsule would require the enhanced production of glutamate, evidence for which is also found in the transcriptome analysis. Glutamate is metabolised from 2-oxoglutarate and glutamine via glytamate synthase (GSU1239) and glutamine is replenished from glutamate and ammonium by the glutamine synthetase (GSU1835, see green arrows in **Figure 3.22**). The genes for both of these enzymes were expressed at up to 3.6-fold higher levels in experiment 4F5%. Additionally, transcript levels of genes encoding for a 2-oxoglutarate ferredoxin oxidoreductase were strongly reduced (up to 52-fold). This enzyme carries out the reaction from 2-oxoglutarate to succinyl-CoA. Thus, carbon flow down the TCA cycle is lowered, permitting the production of glutamate (**Figure 3.22**). Also, GSU2089 is downregulated in cultivation 4F5%. This gene encodes for a rod shape determining protein. Renouncing its typical rod shape would result in more spherical cells, and consequently the surface to volume ratio of *G. sulfurreducens* would be lowered, thus displaying less area for oxygen to enter the cells.

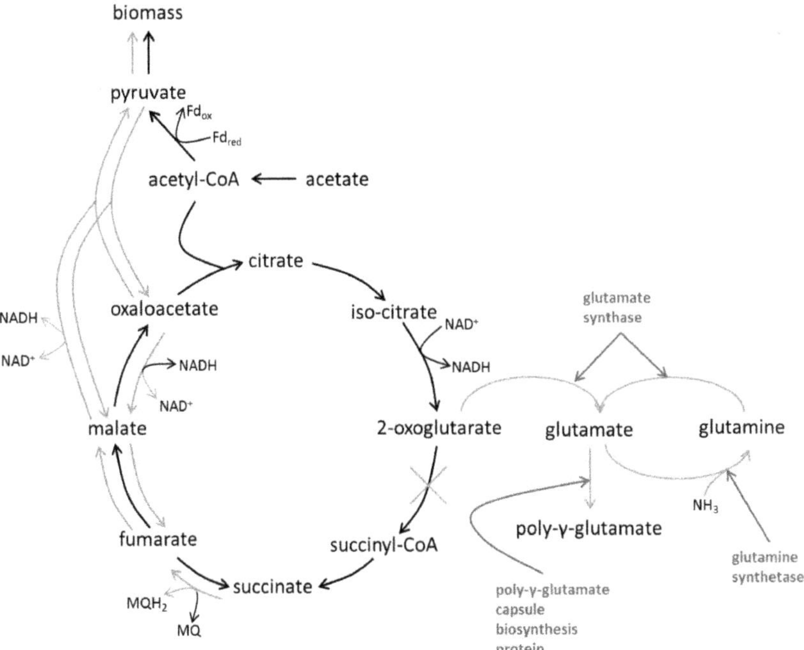

Figure 3.22: Flow of carbon around the tricarboxylic acid (TCA) cycle under fumarate respiring conditions (black arrows) as derived from Butler et al. (2006) and Segura et al. (2008) and proposed alterations in carbon flow during electron acceptor limiting (orange arrows), microaerobic growth supporting (blue arrows) and microaerobic growth inhibiting (green arrows) conditions. Fd = ferrodoxin, MQ = menaquinone, MQH_2 = menaquinol. Red annotations refer to enzymes responsible for the indicated reactions.

An additional indication for biofilm formation in experiment 4F5% is the up to 4-fold upregulation of the expression of GSU0078, GSU3356 and GSU2828. The former gene encodes for a PilZ domain containing protein while the latter both encode for diguanylate cyclase, an enzyme that forms cyclic diguanylate (c-di-GMP) from GTP. c-di-GMP binds to PilZ domains of bacterial pili and thus serves as signalling molecule for biofilm formation and inhibition of motility (Sondermann et al. 2012; Boyd and O'Toole 2012; Romling et al. 2013). Additionally, the type IV pili genes seen expressed at higher levels in cultivation 4F1% are found at lower levels here for the most part (see **Table 3.1**). This implies that *G. sulfurreducens*' strategy when confronted with amounts of oxygen that are too high to allow for complete reduction is encapsulation rather than fleeing or partial reduction.

Transcriptome analysis revealed a lot of indicators as to how *G. sulfurreducens* reacts to oxygen exposure. The higher expression of type IV pili was thought to be either a response to low electron acceptor concentrations or a means of fleeing low oxygen concentrations. To clarify the role of type IV pili, transcriptome analysis of cells grown under fumarate limiting but anaerobic conditions could be conducted. If transcript levels of these genes are still elevated, the former will probably be the case. If, however, transcript levels return to normal values, involvement of the pili in motility becomes even more likely. Another useful technique to apply for studying the functions of the different enzymes discussed here is the construction of deletion mutants. For example, the responsibility of the cyochrome bd menaquinol oxidase for oxygen reduction could be proven by deleting the corresponding gene and investigating whether cell growth with oxygen by *G. sulfurreducens* is still possible.

Since *G. sulfurreducens* is a highly important organism for BES, the possibility of oxygen reduction by cells that grow using anodes as terminal electron acceptor is worthwhile to investigate in future experiments. Short-term exposure to oxygen was shown to impair current generation (Li et al. 2012), most likely due to either a reduction of oxygen instead of the anode or possibly due to complete growth inhibition similar as to what was observed here in experiment 4F5%. Which of the two hypotheses is correct could be investigated by cultivating *G. sulfurreducens* electrochemically and introducing small amounts of oxygen while monitoring the DO concentration. If no reduction of oxygen occurs, current production will likely only be reestablished by external intervention and the restoring of anaerobic conditions. Depending on the time of oxygen exposure, there will likely be cell damage that impairs the maximum current production. If, on the other hand, reduction of oxygen occurs, it could be investigated whether *G. sulfurreducens* eventually restores the maximum current production autonomously. Thus, the long-term effects of oxygen on biofilm formation and current production could be clarified, especially since evidence was seen here that oxygen has an influence on the biofilm formation of *G. sulfurreducens*. This might help to evaluate which measures need to be taken when designing pure culture BES to keep oxygen intrusion at an acceptable level.

4 Conclusion

As important biocatalysts, electrochemically active bacteria need to be studied to understand them on a molecular level and to ultimately improve their performance in BES. Due to the great difference in electrochemical performance between pure and mixed cultures, defined mixed cultures have been investigated to understand the specific interactions between bacteria, however only for short timescales (Dolch et al. 2014; Prokhorova et al. 2017). There is further great discrepancy within literature regarding the choice of anode potential for optimal performance (Babauta et al. 2012a). Lastly, there is little understanding as to how oxygen influences *G. sulfurreducens* within these systems. Apart from the discovery of the general ability of this bacterium to reduce low amounts of oxygen by Lin et al. (2004), almost no continuative research has been performed so far. This factual situation led to the three goals of this thesis.

- Investigating the long-term behaviour of interactions between electrochemically active bacteria in a mixed culture.
- Determining the anode potential that provides the optimal conditions for bacterial performance in MECs.
- Discovering how *G. sulfurreducens* reacts to oxygen intrusion in its living environment.

For the first goal, specific emphasis was laid on a defined mixed culture of *G. sulfurreducens* and *S. oneidensis* with special regards to the incorporation of *S. oneidensis* into the electrochemically active biofilm. Positive effects of *S. oneidensis* on the *G. sulfurreducens* based biofilm could initially be observed here. However, these effects were not associated with an incorporation of *S. oneidensis* into the biofilm but rather with planktonic cells. Thus, the positive effects were not long-term stable if planktonic cells were removed. Therefore, when intending to make use of the positive interactions of mixed cultures, it has to be ensured that the organisms responsible for these positive interactions are kept viable in the system. This could be achieved by operating the MEC in continuous mode rather then exchanging the medium completely when substrate is depleted so that a constant and stable stationary concentration of planktonic cells and mediator will be established. Planktonic cell growth of *S. oneidensis* might further be facilitated by the controlled introduction of small amounts of oxygen into the MEC. An established stable defined mixed culture could be used for improved H_2 production in MEC, as the non-methanogenic bacteria *G. sulfurreducens* and *S. oneidensis* would not lower H_2 yields as opposed to methanogens in undefined mixed cultures (Call et al. 2009).

The second goal was to investigate the influence of the set anode potential on the electron transfer and to find the best operating potential for the defined mixed culture of *G. sulfurreducens* and *S. oneidensis*. The highest current density and biofilm thickness were seen at 0.2 $V_{Ag/AgCl}$. This would result in the highest theoretical H_2 production rate. However, at -0.2 $V_{Ag/AgCl}$, the current density achieved in relation to the biofilm thickness was higher than at 0.2 $V_{Ag/AgCl}$, which suggests that at -0.2 $V_{Ag/AgCl}$ biofilms can be established that are more productive. In future experiments, cells could be conditioned at -0.2 $V_{Ag/AgCl}$ with a later potential switch to 0.2 $V_{Ag/AgCl}$ to achieve a thick but highly productive biofilm, which might lead to improved current density and thus H_2 production rates.

Finally, the third goal of this thesis was to elucidate how *G. sulfurreducens* performs oxygen reduction. It was found that *G. sulfurreducens* has to reduce all oxygen immediately and completely in order to be able to respire it. Failing that, the bacterium will form protective layers to keep oxygen induced cell damage to a minimum rather than comsuming oxygen only partially. The cytochrome bd menaquinol oxidase is the enzyme most likely responsible for oxygen reduction. To prove this hypothesis, the corresponding gene should be knocked out and the cell growth characteristics of the resulting mutant with oxygen should be analysed.

Bringing together the different aspects considered here, the influence of oxygen should also be investigated on cells grown electrochemically in future experiments. Oxygen intrusion in BESs has been shown to impair current production (Li et al. 2012), but the response of *G. sulfurreducens* to such an intrusion has not yet been investigated. There is a great difference between the ceasing of current due to cell death or due to a temporary occupation with oxygen removal. This might have implications for the anaerobic design and operation of BESs in the future.

5 References

All online sources indicated in the text by web-source (WS) are listed at the bottom of this section.

Acworth IN (2003) The handbook of redox biochemistry, 1st edn. ESA, Chelmsford

Aelterman P, Freguia S, Keller J, Verstraete W, Rabaey K (2008) The anode potential regulates bacterial activity in microbial fuel cells. Appl Microbiol Biotechnol 78:409–418 . doi: 10.1007/s00253-007-1327-8

Babauta J, Renslow R, Lewandowski Z, Beyenal H (2012a) Electrochemically active biofilms: Facts and fiction. A review. Biofouling 28:789–812 . doi: 10.1080/08927014.2012.710324

Babauta JT, Nguyen HD, Harrington TD, Renslow R, Beyenal H (2012b) pH, redox potential and local biofilm potential microenvironments within *Geobacter sulfurreducens* biofilms and their roles in electron transfer. Biotechnol Bioeng 109:2651–2662 . doi: 10.1002/bit.24538

Balch WE, Fox GE, Magrum LJ, Woese CR, Wolfe RS (1979) Methanogens: reevaluation of a unique biological group. Microbiol Rev 43:260–296 . doi: 10.1016/j.watres.2010.10.010

Baudler A, Schmidt I, Langner M, Greiner A, Schröder U (2015) Does it have to be carbon? Metal anodes in microbial fuel cells and related bioelectrochemical systems. Energy Environ Sci 8:2048–2055 . doi: 10.1039/C5EE00866B

Bertani G (1951) Studies on lysogenesis. I. The mode of phage liberation by lysogenic *Escherichia coli*. J Bacteriol 62:293–300

Biedendieck R, Borgmeier C, Bunk B, Stammen S, Scherling C, Meinhardt F, Wittmann C, Jahn D (2011) Systems biology of recombinant protein production using *Bacillus megaterium*, 1st edn. Elsevier Inc., Oxford

Binnenkade L, Teichmann L, Thormanna KM (2014) Iron triggers λSo prophage induction and release of extracellular DNA in *Shewanella oneidensis* MR-1 biofilms. Appl Environ Microbiol 80:5304–5316 . doi: 10.1128/AEM.01480-14

Binnewies M, Jäckel M, Willner H, Rayner-Canham G (2004) Allgemeine und Anorganische Chemie, 1st edn. Spektrum Akademischer Verlag, Heidelberg

Block J (2018) Sauerstoffeinfluss auf die Stromproduktion elektrochemisch aktiver Mischkultur-Biofilme in mikrobiellen Elektrolysezellen. Bachelorarbeit, Technische Universität Braunschweig

Bond DR, Lovley DR (2003) Electricity production by *Geobacter sulfurreducens* attached to electrodes. Appl Environ Microbiol 69:1548–1555 . doi: 10.1128/AEM.69.3.1548

Bond DR, Strycharz-Glaven SM, Tender LM, Torres CI (2012) On electron transport through *Geobacter* biofilms. ChemSusChem 5:1099–1105 . doi: 10.1002/cssc.201100748

Borgmeier C, Biedendieck R, Hoffmann K, Jahn D, Meinhardt F (2011) Transcriptome profiling of degU expression reveals unexpected regulatory patterns in *Bacillus megaterium* and discloses new targets for optimizing expression. Appl Microbiol Biotechnol 92:583–596 . doi: 10.1007/s00253-011-3575-x

Borole AP, Reguera G, Ringeisen B, Wang Z, Feng Y, Kim BH (2011) Electroactive biofilms: Current status and future research needs. Energy Environ Sci 4:4813–4834 . doi: 10.1039/c1ee02511b

Bosch J, Lee KY, Hong SF, Harnisch F, Schröder U, Meckenstock RU (2014) Metabolic efficiency of *Geobacter sulfurreducens* growing on anodes with different redox potentials. Curr Microbiol 68:763–768 . doi: 10.1007/s00284-014-0539-2

Boyd CD, O'Toole GA (2012) Second messenger regulation of biofilm formation: Breakthroughs in understanding c-di-GMP effector Systems. Annu Rev Cell Dev Biol 28:439–462 . doi: 10.1146/annurev-cellbio-101011-155705

Breuer M, Rosso KM, Blumberger J, Butt JN (2015) Multi-haem cytochromes in *Shewanella oneidensis* MR-1: Structures, functions and opportunities. J R Soc Interface 12:20141117 . doi: 10.1098/rsif.2014.1117

Brown DG, Komlos J, Jaffé PR (2005) Simultaneous utilization of acetate and hydrogen by *Geobacter sulfurreducens* and implications for use of hydrogen as an indicator of redox conditions. Environ Sci Technol 39:3069–3076 . doi: 10.1021/es048613p

Burrows LL (2012) *Pseudomonas aeruginosa* Twitching motility: Type IV pili in action. Annu Rev Microbiol 66:493–520 . doi: 10.1146/annurev-micro-092611-150055

Busalmen JP, Esteve-Nuñez A, Feliu JM (2008) Whole cell electrochemistry of electricity-producing microorganisms evidence an adaptation for optimal exocellular electron transport. Environ Sci Technol 42:2445–2450 . doi: 10.1021/es702569y

Butler JE, Glaven RH, Esteve-Núnez A, Nunez C, Shelobolina ES, Bond DR, Lovley DR (2006) Genetic characterization of a single bifunctional enzyme for fumarate reduction and succinate oxidation in *Geobacter sulfurreducens* and engineering of fumarate reduction in *Geobacter metallireducens*. Microbiology 188:450–455 . doi: 10.1128/JB.188.2.450

Caccavo F, Lonergan DJ, Lovley DR, Davis M, Stolz JF, McInerney MJ (1994) *Geobacter sulfurreducens* sp. nov., a hydrogen- and acetate-oxidizing dissimilatory metal-reducing microorganism. Appl Environ Microbiol 60:3752–3759 . doi: 0099-2240/$04.00+0

Call DF, Logan BE (2011) Lactate oxidation coupled to iron or electrode reduction by *Geobacter sulfurreducens* PCA. Appl Environ Microbiol 77:8791–8794 . doi: 10.1128/AEM.06434-11

Call DF, Wagner RC, Logan BE (2009) Hydrogen production by *Geobacter* species and a mixed consortium in a microbial electrolysis cell. Appl Environ Microbiol 75:7579–7587 . doi: 10.1128/AEM.01760-09

Carmona-Martinez AA, Harnisch F, Fitzgerald LA, Biffinger JC, Ringeisen BR, Schröder U (2011) Cyclic voltammetric analysis of the electron transfer of *Shewanella oneidensis* MR-1 and nanofilament and cytochrome knock-out mutants. Bioelectrochemistry 81:74–80 . doi: 10.1016/j.bioelechem.2011.02.006

Coursolle D, Gralnick JA (2012) Reconstruction of extracellular respiratory pathways for iron(III) reduction in *Shewanella oneidensis* strain MR-1. Front Microbiol 3:1–11 . doi: 10.3389/fmicb.2012.00056

Daum B, Gold V (2018) Twitch or swim: Towards the understanding of prokaryotic motion based on the type IV pilus blueprint. Biol Chem 399:799–808 . doi: 10.1515/hsz-2018-0157

de Bolster MWG (1997) Glossary of terms used in bioinorganic chemistry. Pure Appl Chem 69:1251–1304 . doi: 10.1351/pac199769061251

DiDonato LN, Sullivan SA, Methé BA, Nevin KP, England R, Lovley DR (2006) Role of RelGsu in stress response and Fe(III) reduction in *Geobacter sulfurreducens*. J Bacteriol 188:8469–8478 . doi: 10.1128/JB.01278-06

Dolch K, Danzer J, Kabbeck T, Bierer B, Erben J, Förster AH, Maisch J, Nick P, Kerzenmacher S, Gescher J (2014) Characterization of microbial current production as a function of microbe-electrode-interaction. Bioresour Technol 157:284–292 . doi: 10.1016/j.biortech.2014.01.112

Du Z, Li H, Gu T (2007) A state of the art review on microbial fuel cells: A promising technology for wastewater treatment and bioenergy. Biotechnol Adv 25:464–482 . doi: 10.1016/j.biotechadv.2007.05.004

Esteve-Núñez A, Núñez C, Lovley DR (2004) Preferential reduction of Fe (III) over fumarate by *Geobacter sulfurreducens*. J Bacteriol 186:2897–2899 . doi: 10.1128/JB.186.9.2897

Esteve-Núñez A, Rothermich M, Sharma M, Lovley D (2005) Growth of *Geobacter sulfurreducens* under nutrient-limiting conditions in continuous culture. Environ Microbiol 7:641–648 . doi: 10.1111/j.1462-2920.2005.00731.x

Finkelstein DA, Tender LM, Zeikus JG (2006) Effect of electrode potential on electrode-reducing microbiota. Environ Sci Technol 40:6990–6995 . doi: 10.1021/es061146m

Franks AE, Nevin KP, Glaven RH, Lovley DR (2010) Microtoming coupled to microarray analysis to evaluate the spatial metabolic status of *Geobacter sulfurreducens* biofilms. ISME J 4:509–519 . doi: 10.1038/ismej.2009.137

Fricke K, Harnisch F, Schröder U (2008) On the use of cyclic voltammetry for the study of anodic electron transfer in microbial fuel cells. Energy Environ Sci 1:144 . doi: 10.1039/b802363h

Galushko AS, Schink B (2000) Oxidation of acetate through reactions of the citric acid cycle by *Geobacter sulfurreducens* in pure culture and in syntrophic coculture. Arch Microbiol 174:314–321 . doi: 10.1007/s002030000208

Gödeke J, Paul K, Lassak J, Thormann KM (2011) Phage-induced lysis enhances biofilm formation in *Shewanella oneidensis* MR-1. ISME J 5:613–626 . doi: 10.1038/ismej.2010.153

Gorby YA, Yanina S, McLean JS, Rosso KM, Moyles D, Dohnalkova A, Beveridge TJ, Chang IS, Kim BH, Kim KS, Culley DE, Reed SB, Romine MF, Saffarini DA, Hill EA, Shi L, Elias DA, Kennedy DW, Pinchuk GE, Watanabe K, Ishii S, Logan BE, Nealson KH, Fredrickson JK (2006) Electrically conductive bacterial nanowires produced by *Shewanella oneidensis* strain MR-1 and other microorganisms. Proc Natl Acad Sci 106:11358–11636 . doi: 10.1073/pnas.0604517103

Grobbler C, Virdis B, Nouwens A, Harnisch F, Rabaey K, Bond PL (2018) Effect of the anode potential on the physiology and proteome of *Shewanella oneidensis* MR-1. Bioelectrochemistry 119:172–179 . doi: 10.1016/j.bioelechem.2017.10.001

Gu Y, Li Y, Li X, Luo P, Wang H, Robinson ZP, Wang X, Wu J (2017) The feasibility and challenges of energy self-sufficient wastewater treatment plants. Appl Energy 204:1463–1475 . doi: 10.1016/j.apenergy.2017.02.069

Hallenbeck PC, Grogger M, Veverka D (2014) Recent advances in microbial electrocatalysis. Electrocatalysis 5:319–329 . doi: 10.1007/s12678-014-0198-x

Hasford JJ, Kemnitzer W, Rizzo CJ (1997) Conformational effects on flavin redox chemistry. J Org Chem 62:5244–5245 . doi: 10.1021/jo9703865

Hau HH, Gilbert A, Coursolle D, Gralnick JA (2008) Mechanism and consequences of anaerobic respiration of cobalt by *Shewanella oneidensis* strain MR-1. Appl Environ Microbiol 74:6880–6886 . doi: 10.1128/AEM.00840-08

Hell S, Reiner G, Cremer C, Stelzer EHK (1993) Aberrations in confocal fluorescence microscopy induced by mismatches in refractive index. J Microsc 169:391–405 . doi: 10.1111/j.1365-2818.1993.tb03315.x

Hobbie JE, Daley RJ, Jasper S (1977) Use of nuclepore filter counting bacteria by fluoroscence microscopy. Appl Environ Microbiol Microbiol 33:1225–1228 . doi: citeulike-article-id:4408959

Holmes DE, Chaudhuri SK, Nevin KP, Mehta T, Methé BA, Liu A, Ward JE, Woodard TL, Webster J, Lovley DR (2006) Microarray and genetic analysis of electron transfer to electrodes in *Geobacter sulfurreducens*. Environ Microbiol 8:1805–1815 . doi: 10.1111/j.1462-2920.2006.01065.x

Inoue K, Leang C, Franks AE, Woodard TL, Nevin KP, Lovley DR (2011) Specific localization of the c-type cytochrome OmcZ at the anode surface in current-producing biofilms of *Geobacter sulfurreducens*. Environ Microbiol Rep 3:211–217 . doi: 10.1111/j.1758-2229.2010.00210.x

Ishii S, Watanabe K, Yabuki S, Logan BE, Sekiguchi Y (2008) Comparison of electrode reduction activities of *Geobacter sulfurreducens* and an enriched consortium in an air-cathode microbial fuel cell. Appl Environ Microbiol 74:7348–7355 . doi: 10.1128/AEM.01639-08

Jenney Jr. FE (1999) Anaerobic microbes: Oxygen detoxification without superoxide dismutase. Science (80-) 286:306–309 . doi: 10.1126/science.286.5438.306

Kabashima Y, Kishikawa JI, Kurokawa T, Sakamoto J (2009) Correlation between proton translocation and growth: Genetic analysis of the respiratory chain of *Corynebacterium glutamicum*. J Biochem 146:845–855 . doi: 10.1093/jb/mvp140

Kim JR, Min B, Logan BE (2005) Evaluation of procedures to acclimate a microbial fuel cell for electricity production. Appl Microbiol Biotechnol 68:23–30 . doi: 10.1007/s00253-004-1845-6

Koch C, Günther S, Desta AF, Hübschmann T, Müller S (2013) Cytometric fingerprinting for analyzing microbial intracommunity structure variation and identifying subcommunity function. Nat Protoc 8:190–202 . doi: 10.1038/nprot.2012.149

Koch C, Harnisch F, Schröder U, Müller S (2014) Cytometric fingerprints: Evaluation of new tools for analyzing microbial community dynamics. Front Microbiol 5:273 . doi: 10.3389/fmicb.2014.00273

Kotloski NJ, Gralnick JA (2013) Flavin electron shuttles dominate extracellular electron transfer by *Shewanella oneidensis*. MBio 4:10–13 . doi: 10.1128/mBio.00553-12

Krämer CEM, Wiechert W, Kohlheyer D (2016) Time-resolved, single-cell analysis of induced and programmed cell death via non-invasive propidium iodide and counterstain perfusion. Sci Rep 6:1–13 . doi: 10.1038/srep32104

Kumar A, Siggins A, Katuri K, Mahony T, O'Flaherty V, Lens P, Leech D (2013) Catalytic response of microbial biofilms grown under fixed anode potentials depends on electrochemical cell configuration. Chem Eng J 230:532–536 . doi: 10.1016/j.cej.2013.06.044

Leang C, Coppi M V, Lovley DR (2003) OmcB, a c-type polyheme cytochrome, involved in Fe(III) reduction in *Geobacter sulfurreducens*. Society 185:2096–2103 . doi: 10.1128/JB.185.7.2096

Leang C, Malvankar NS, Franks AE, Nevin KP, Lovley DR (2013) Engineering *Geobacter sulfurreducens* to produce a highly cohesive conductive matrix with enhanced capacity for current production. Energy Environ Sci 6:1901 . doi: 10.1039/c3ee40441b

Leang C, Qian X, Mester T, Lovley DR (2010) Alignment of the c-type cytochrome OmcS along pili of *Geobacter sulfurreducens*. Appl Environ Microbiol 76:4080–4084 . doi: 10.1128/AEM.00023-10

Levar CE, Chan CH, Mehta-kolte MG (2014) An inner membrane cytochrome required only for reduction of high redox potential extracellular electron acceptors. MBio 5:02034-14 . doi: 10.1128/mBio.02034-14

Li Z, Venkataraman A, Rosenbaum MA, Angenent LT (2012) A laminar-flow microfluidic device for quantitative analysis of microbial electrochemical activity. ChemSusChem 5:1119–1123 . doi: 10.1002/cssc.201100736

Lin WC, Coppi M V, Lovley DR, Lovley DR (2004) *Geobacter sulfurreducens* can grow with oxygen as a terminal electron acceptor. Appl Environ Microbiol 70:2525–2528 . doi: 10.1128/AEM.70.4.2525

Liu J, Hou H, Chen X, Bazan GC, Kashima H, Logan BE (2015) Conjugated oligoelectrolyte represses hydrogen oxidation by *Geobacter sulfurreducens* in microbial electrolysis cells. Bioelectrochemistry 106:379–382 . doi: 10.1016/j.bioelechem.2015.07.001

Liu Y, Bond DR (2012) Long-distance electron transfer by *G. sulfurreducens* biofilms results in accumulation of reduced c-type cytochromes. ChemSusChem 5:1047–1053 . doi: 10.1002/cssc.201100734

Liu Y, Wang Z, Liu J, Levar C, Edwards MJ, Babauta JT, Kennedy DW, Shi Z, Beyenal H, Bond DR, Clarke TA, Butt JN, Richardson DJ, Rosso KM, Zachara JM, Fredrickson JK, Shi L (2014) A trans-outer membrane porin-cytochrome protein complex for extracellular electron transfer by *Geobacter sulfurreducens* PCA. Environ Microbiol Rep 6:776–785 . doi: 10.1111/1758-2229.12204

Logan BE (2009) Exoelectrogenic bacteria that power microbial fuel cells. Nat Rev Microbiol 7:375–381 . doi: 10.1038/nrmicro2113

Logan BE, Call D, Cheng S, Hamelers HVM, Sleutels THJA, Jeremiasse AW, Rozendal RA (2008) Microbial electrolysis cells for high yield hydrogen gas production from organic matter. Environ Sci Technol 42:8630–8640 . doi: 10.1021/es801553z

Lombard M, Fontecave M, Touati D, Nivière V (2000) Reaction of the desulfoferrodoxin from *Desulfoarculus baarsii* with superoxide anion. J Biol Chem 275:115–121 . doi: 10.1074/jbc.275.1.115

Lovley DR (2006) Bug juice: Harvesting electricity with microorganisms. Nat Rev Microbiol 4:497–508 . doi: 10.1038/nrmicro1442

Lovley DR (2017) Syntrophy goes electric: Direct interspecies electron transfer. Annu Rev Microbiol 71: . doi: 10.1146/annurev-micro-030117-020420

Lovley DR, Holmes DE, Nevin KP (2004) Dissimilatory Fe(III) and Mn(IV) reduction. Adv Microb Physiol 49:219–286 . doi: 10.1016/S0065-2911(04)49005-5

Lovley DR, Malvankar NS (2015) Seeing is believing: novel imaging techniques help clarify microbial nanowire structure and function. Environ Microbiol 17:2209–2215 . doi: 10.1111/1462-2920.12708

Mahadevan R, Bond DR, Butler JE, Coppi V, Palsson BO, Schilling CH, Lovley DR (2006) Characterization of metabolism in the Fe(III)-reducing organism *Geobacter sulfurreducens* by constraint-based modeling. Appl Environ Microbiol 72:1558–1568 . doi: 10.1128/AEM.72.2.1558

Malvankar NS, Lovley DR (2014) Microbial nanowires for bioenergy applications. Curr Opin Biotechnol 27:88–95 . doi: 10.1016/j.copbio.2013.12.003

Malvankar NS, Tuominen MT, Lovley DR (2012) Biofilm conductivity is a decisive variable for high-current-density *Geobacter sulfurreducens* microbial fuel cells. Energy Environ Sci 5:5790 . doi: 10.1039/c2ee03388g

Marsili E, Baron DB, Shikhare ID, Coursolle D, Gralnick JA, Bond DR (2008a) *Shewanella* secretes flavins that mediate extracellular electron transfer. Proc Natl Acad Sci 105:3968–3973 . doi: 10.1073/pnas.0710525105

Marsili, Rollefson JB, Baron DB, Hozalski RM, Bond DR (2008b) Microbial biofilm voltammetry: Direct electrochemical characterization of catalytic electrode-attached biofilms. Appl Environ Microbiol 74:7329–7337 . doi: 10.1128/AEM.00177-08

Mehta T, Coppi M V, Childers SE, Lovley DR (2005) Outer membrane c-type cytochromes required for Fe(III) and Mn(IV) oxide reduction in *Geobacter sulfurreducens*. Microbiology 71:8634–8641 . doi: 10.1128/AEM.71.12.8634

Meitl LA, Eggleston CM, Colberg PJS, Khare N, Reardon CL, Shi L (2009) Electrochemical interaction of *Shewanella oneidensis* MR-1 and its outer membrane cytochromes OmcA and MtrC with hematite electrodes. Geochim Cosmochim Acta 73:5292–5307 . doi: 10.1016/j.gca.2009.06.021

Meshulam-Simon G, Behrens S, Choo AD, Spormann AM (2007) Hydrogen metabolism in *Shewanella oneidensis* MR-1. Appl Environ Microbiol 73:1153–1165 . doi: 10.1128/AEM.01588-06

Methé BA, Nelson KE, Eisen JA, Paulsen IT, Nelson W, Heidelberg JF, Wu D, Wu M, Ward N, Beanan MJ, Dodson RJ, Madupu R, Brinkac LM, Daugherty SC, DeBoy RT, Durkin AS, Gwinn M, Kolonay JF, Sullivan SA, Haft DH, Selengut J, Davidsen TM, Zafar N, White O, Tran B, Romero C, Forberger HA, Weidman J, Khouri H, Feldblyum T V, Utterback TR, Van Aken SE, Lovley DR, Fraser CM (2003) Genome of *Geobacter sulfurreducens*: Metal reduction in subsurface environments. Science (80-) 302:1967–1969 . doi: 10.1126/science.1088727

Mouser PJ, Holmes DE, Perpetua LA, DiDonato R, Postier B, Liu A, Lovley DR (2009) Quantifying expression of *Geobacter* spp. oxidative stress genes in pure culture and during in situ uranium bioremediation. ISME J 3:454–465 . doi: 10.1038/ismej.2008.126

Müller S (2007) Modes of cytometric bacterial DNA pattern: A tool for pursuing growth. Cell Prolif 40:621–639 . doi: 10.1111/j.1365-2184.2007.00465.x

Myers CR, Nealson KH (1988) Bacterial manganese reduction and growth with manganese oxide as the sole electron acceptor. Science (80-) 240:1319–1321 . doi: 10.1126/science.240.4857.1319

Nelson DL, Cox MM (2017) Lehninger Principles of biochemistry. Macmillan Education, London

Nevin KP, Kim BC, Glaven RH, Johnson JP, Woodward TL, Methé BA, Didonato RJ, Covalla SF, Franks AE, Liu A, Lovley DR (2009) Anode biofilm transcriptomics reveals outer surface components essential for high density current production in *Geobacter sulfurreducens* fuel cells. PLoS One 4:e5628 . doi: 10.1371/journal.pone.0005628

Nevin KP, Richter H, Covalla SF, Johnson JP, Woodard TL, Orloff AL, Jia H, Zhang M, Lovley DR (2008) Power output and columbic efficiencies from biofilms of *Geobacter sulfurreducens* comparable to mixed community microbial fuel cells. Environ Microbiol 10:2505–2514 . doi: 10.1111/j.1462-2920.2008.01675.x

Nunez C, Esteve-Nunez A, Giometti C, Tollaksen S, Khare T, Lin W, Lovley DR, Methe BA (2006) DNA Microarray and Proteomic Analyses of the RpoS Regulon in *Geobacter sulfurreducens*. J Bacteriol 188:2792–2800 . doi: 10.1128/JB.188.8.2792-2800.2006

Okamoto A, Hashimoto K, Nealson KH, Nakamura R (2013) Rate enhancement of bacterial extracellular electron transport involves bound flavin semiquinones. Proc Natl Acad Sci 110:7856–7861 . doi: 10.1073/pnas.1220823110

Okamoto A, Saito K, Inoue K, Nealson KH, Hashimoto K, Nakamura R (2014) Uptake of self-secreted flavins as bound cofactors for extracellular electron transfer in *Geobacter* species. Energy Environ Sci 7:1357–1361 . doi: 10.1039/C3EE43674H

Ordóñez M V., Schrott GD, Massazza DA, Busalmen JP (2016) The relay network of *Geobacter* biofilms. Energy Environ Sci 9:2677–2681 . doi: 10.1039/c6ee01699e

Pinchuk GE, Geydebrekht O V., Hill EA, Reed JL, Konopka AE, Beliaev AS, Fredrickson JK (2011) Pyruvate and lactate metabolism by *Shewanella oneidensis* MR-1 under fermentation, oxygen limitation, and fumarate respiration conditions. Appl Environ Microbiol 77:8234–8240 . doi: 10.1128/AEM.05382-11

Pirbadian S, Barchinger SE, Leung KM, Byun HS, Jangir Y, Bouhenni RA, Reed SB, Romine MF, Saffarini DA, Shi L, Gorby YA, Golbeck JH, El-Naggar MY (2014) *Shewanella oneidensis* MR-1 nanowires are outer membrane and periplasmic extensions of the extracellular electron transport components. Proc Natl Acad Sci 111:12883–12888 . doi: 10.1073/pnas.1410551111

Prokhorova A, Sturm-Richter K, Doetsch A, Gescher J (2017) Resilience, dynamics, and interactions within a model multispecies exoelectrogenic-biofilm community. Appl Environ Microbiol 83:1–15 . doi: 10.1128/AEM.03033-16

Reguera G (2018) Microbial nanowires and electroactive biofilms. FEMS Microbiol Ecol 94:1–13 . doi: 10.1093/femsec/fiy086

Reguera G, McCarthy KD, Mehta T, Nicoll JS, Tuominen MT, Lovley DR (2005) Extracellular electron transfer via microbial nanowires. Nature 435:1098–1101 . doi: 10.1038/nature03661

Renslow RS, Babauta JT, Dohnalkova AC, Boyanov MI, Kemner KM, Majors PD, Fredrickson JK, Beyenal H (2013) Metabolic spatial variability in electrode-respiring *Geobacter sulfurreducens* biofilms. Energy Environ Sci 6:1827–1836 . doi: 10.1039/c3ee40203g

Richter L V., Sandler SJ, Weis RM (2012) Two isoforms of *Geobacter sulfurreducens* PilA have distinct roles in pilus biogenesis, cytochrome localization, extracellular electron transfer, and biofilm formation. J Bacteriol 194:2551–2563 . doi: 10.1128/JB.06366-11

Romling U, Galperin MY, Gomelsky M (2013) Cyclic di-GMP: the first 25 years of a universal bacterial second messenger. Microbiol Mol Biol Rev 77:1–52 . doi: 10.1128/MMBR.00043-12

Rosa LFM, Hunger S, Gimkiewicz C, Zehnsdorf A, Harnisch F (2017) Paving the way for bioelectrotechnology: Integrating electrochemistry into bioreactors. Eng Life Sci 17:77–85 . doi: 10.1002/elsc.201600105

Rosenbaum M, Cotta MA, Angenent LT (2010) Aerated *Shewanella oneidensis* in continuously fed bioelectrochemical systems for power and hydrogen production. Biotechnol Bioeng 105:880–888 . doi: 10.1002/bit.22621

Rosenbaum MA, Bar HY, Beg QK, Segrè D, Booth J, Cotta MA, Angenent LT (2011) *Shewanella oneidensis* in a lactate-fed pure-culture and a glucose-fed co-culture with *Lactococcus lactis* with an electrode as electron acceptor. Bioresour Technol 102:2623–2628 . doi: 10.1016/j.biortech.2010.10.033

Rosenbaum MA, Franks AE (2014) Microbial catalysis in bioelectrochemical technologies: Status quo, challenges and perspectives. Appl Microbiol Biotechnol 98:509–518 . doi: 10.1007/s00253-013-5396-6

Santos TC, Silva MA, Morgado L, Dantas JM, Salgueiro CA (2015) Diving into the redox properties of *Geobacter sulfurreducens* cytochromes: A model for extracellular electron transfer. 9335–9344 . doi: 10.1039/c5dt00556f

Schmidt I, Pieper A, Wichmann H, Bunk B, Huber K, Overmann J, Walla PJ, Schröder U (2017) In situ autofluorescence spectroelectrochemistry for the study of microbial extracellular electron transfer. ChemElectroChem 4:2515–2519 . doi: 10.1002/celc.201700675

Schroeder RR, Shain I (1969) The application of feedback principles to instrumentation for potentiostatic studies. Instrum Sci Technol 1:233–259 . doi: 10.1080/10739146908543251

Segura D, Mahadevan R, Juárez K, Lovley DR (2008) Computational and Experimental Analysis of Redundancy in the Central Metabolism of *Geobacter sulfurreducens*. PLoS Comput Biol 4:e36 . doi: 10.1371/journal.pcbi.0040036

Shelobolina ES, Coppi M V, Korenevsky AA, DiDonato LN, Sullivan SA, Konishi H, Xu H, Leang C, Butler JE, Kim B-C, Lovley DR (2007) Importance of c-type cytochromes for U(VI) reduction by *Geobacter sulfurreducens*. BMC Microbiol 7:16 . doi: 10.1186/1471-2180-7-16

Sondermann H, Shikuma NJ, Yildiz FH (2012) You've come a long way: c-di-GMP signaling. Curr Opin Microbiol 15:140–146 . doi: 10.1016/j.mib.2011.12.008

Speers AM, Reguera G (2012) Electron donors supporting growth and electroactivity of *Geobacter sulfurreducens* anode biofilms. Appl Environ Microbiol 78:437–444 . doi: 10.1128/AEM.06782-11

Steidl RJ, Lampa-Pastirk S, Reguera G (2016) Mechanistic stratification in electroactive biofilms of *Geobacter sulfurreducens* mediated by pilus nanowires. Nat Commun 7:12217 . doi: 10.1038/ncomms12217

Stephen CS, LaBelle E V., Brantley SL, Bond DR (2014) Abundance of the Multiheme c-Type Cytochrome OmcB Increases in Outer Biofilm Layers of Electrode-Grown *Geobacter sulfurreducens*. PLoS One 9:e104336 . doi: 10.1371/journal.pone.0104336

Summers ZM, Ueki T, Ismail W, Haveman SA, Lovley DR (2012) Laboratory evolution of *Geobacter sulfurreducens* for enhanced growth on lactate via a single-base-pair substitution in a transcriptional regulator. ISME J 6:975–983 . doi: 10.1038/ismej.2011.166

Teravest MA, Angenent LT (2014) Oxidizing electrode potentials decrease current production and coulombic efficiency through cytochromec inactivation in *Shewanella oneidensis* MR-1. ChemElectroChem 1:2000–2006 . doi: 10.1002/celc.201402128

Torres CI, Krajmalnik-Brown R, Parameswaran P, Marcus AK, Wanger G, Gorby YA, Rittmann BE (2009) Selecting anode-respiring bacteria based on anode potential: Phylogenetic, electrochemical, and microscopic characterization. Environ Sci Technol 43:9519–9524 . doi: 10.1021/es902165y

Tran HT, Krushkal J, Antommattei FM, Lovley DR, Weis RM (2008) Comparative genomics of *Geobacter* chemotaxis genes reveals diverse signaling function. BMC Genomics 9:1–15 . doi: 10.1186/1471-2164-9-471

Tremblay PL, Lovley DR (2012) Role of the NiFe hydrogenase hya in oxidative stress defense in *Geobacter sulfurreducens*. J Bacteriol 194:2248–2253 . doi: 10.1128/JB.00044-12

Upmann M (2018) Verbesserung des Elektronentransfers von *Shewanella oneidensis* auf Anoden mittels Fixierung durch Ca-Alginat. Bachelorarbeit, Technische Universität Braunschweig

Venkateswaran K, Moser DP, Dollhopf ME, Lies DP, Saffarini DA, MacGregor BJ, Ringelberg DB, White DC, Nishijima M, Sano H, Burghardt J, Stackebrandt E, Nealson KH (1999) Polyphasic taxonomy of the genus *Shewanella* and description of *Shewanella oneidensis* sp. nov. Int J Syst Bacteriol 49:705–724 . doi: 10.1099/00207713-49-2-705

Wagner RC, Call DF, Logan BE (2010) Optimal set anode potentials vary in bioelectrochemical systems. Environ Sci Technol 44:6036–6041 . doi: 10.1021/es101013e

Wang X, Feng Y, Liu J, Lee H, Ren N (2011) Performance of a batch two-chambered microbial fuel cell operated at different anode potentials. J Chem Technol Biotechnol 86:590–594 . doi: 10.1002/jctb.2558

Wei J, Liang P, Cao X, Huang X (2010) A new insight into potential regulation on growth and power generation of *Geobacter sulfurreducens* in microbial fuel cells based on energy viewpoint. Environ Sci Technol 44:3187–3191 . doi: 10.1021/es903758m

Weisenberger S, Schumpe A (1996) Estimation of gas solublility in salt solution at temperatures from 273 to 363K. AIChE J 42:298–300 . doi: 10.1002/aic.690420130

Wood PM (1988) The potential diagram for oxygen at pH 7. Biochem J 253:287–289 . doi: 10.1042/bj2530287

Zacharoff L, Chan CH, Bond DR (2016) Reduction of low potential electron acceptors requires the CbcL inner membrane cytochrome of *Geobacter sulfurreducens*. Bioelectrochemistry 107:7–13 . doi: 10.1016/j.bioelechem.2015.08.003

Zhu X, Yates MD, Hatzell MC, Rao HA, Saikaly PE, Logan BE (2014) Microbial community composition is unaffected by anode potential. Environ Sci Technol 48:1352–1358 . doi: 10.1021/es501982m.(2)

Zhu X, Yates MD, Logan BE (2012) Set potential regulation reveals additional oxidation peaks of *Geobacter sulfurreducens* anodic biofilms. Electrochem commun 22:116–119 . doi: 10.1016/j.elecom.2012.06.013

Zimmermann J, Hübschmann T, Schattenberg F, Schumann J, Durek P, Riedel R, Friedrich M, Glauben R, Siegmund B, Radbruch A, Müller S, Chang HD (2016) High-resolution microbiota flow cytometry reveals dynamic colitis-associated changes in fecal bacterial composition. Eur J Immunol 46:1300–1303 . doi: 10.1002/eji.201646297

Online Sources

WS-1: https://www.umweltbundesamt.de/daten/wasser/wasserwirtschaft/oeffentliche-abwasserentsorgung#textpart-1 (09.12.2018)

WS-2: https://www.umweltbundesamt.de/sites/default/files/medien/publikation/long/3855.pdf (09.12.2018)

WS-3: https://www.bp.com/content/dam/bp/en/corporate/pdf/energy-economics/statistical-review/bp-stats-review-2018-full-report.pdf (09.12.2018)

6 Appendix

Chronoamperograms and substrate concentration of defined mixed cultures at varying anode potentials:

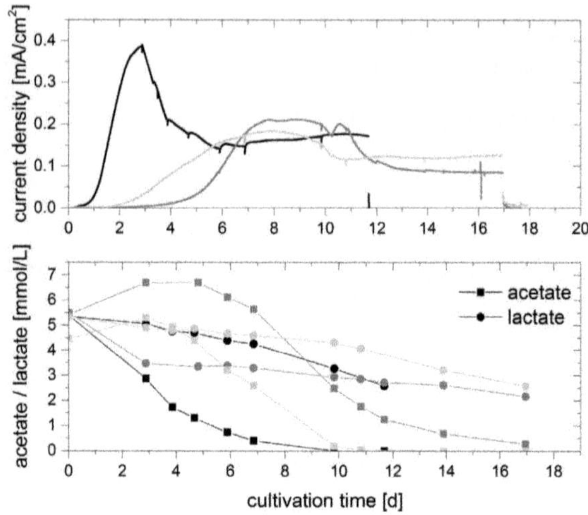

Figure A.1: Chronoamperograms and substrate concentrations of cultivation **cycle 1** of defined mixed cultures of *G. sulfurreducens* and *S. oneidensis* grown at 30 °C on graphite electrodes poised to **-0.2** $V_{Ag/AgCl}$ with 5 mM of acetate and 5 mM of lactate. Differently coloured lines represent individual data from biological triplicates. Spikes in chronoamperogram are due to sampling.

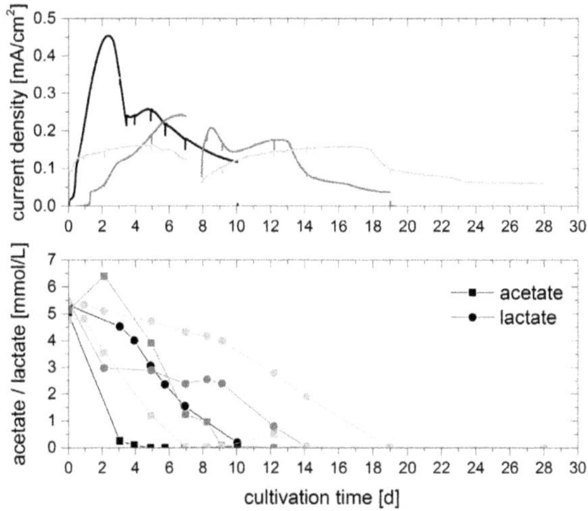

Figure A.2: Chronoamperograms and substrate concentrations of cultivation **cycle 2** of defined mixed cultures of *G. sulfurreducens* and *S. oneidensis* grown at 30 °C on graphite electrodes poised to **-0.2 $V_{Ag/AgCl}$** with 5 mM of acetate and 5 mM of lactate. Differently coloured lines represent individual data from biological triplicates. Spikes in chronoamperogram are due to sampling.

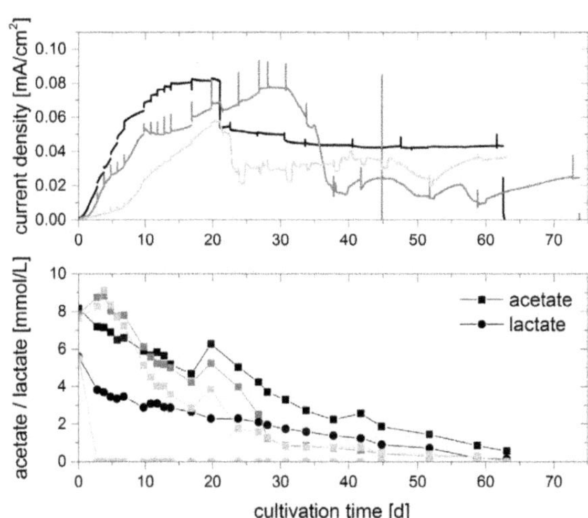

Figure A.1: Chronoamperograms and substrate concentrations of cultivation **cycle 1** of defined mixed cultures of *G. sulfurreducens* and *S. oneidensis* grown at 30 °C on graphite electrodes poised to **0.0 $V_{Ag/AgCl}$** with 5 mM of acetate and 5 mM of lactate. Differently coloured lines represent individual data from biological triplicates. Spikes in chronoamperogram are due to sampling.

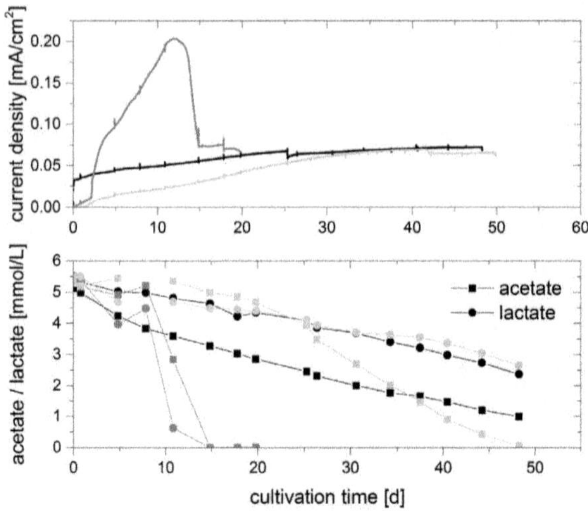

Figure A.4: Chronoamperograms and substrate concentrations of cultivation **cycle 2** of defined mixed cultures of *G. sulfurreducens* and *S. oneidensis* grown at 30 °C on graphite electrodes poised to **0.0** $\mathbf{V_{Ag/AgCl}}$ with 5 mM of acetate and 5 mM of lactate. Differently coloured lines represent individual data from biological triplicates. Spikes in chronoamperogram are due to sampling.

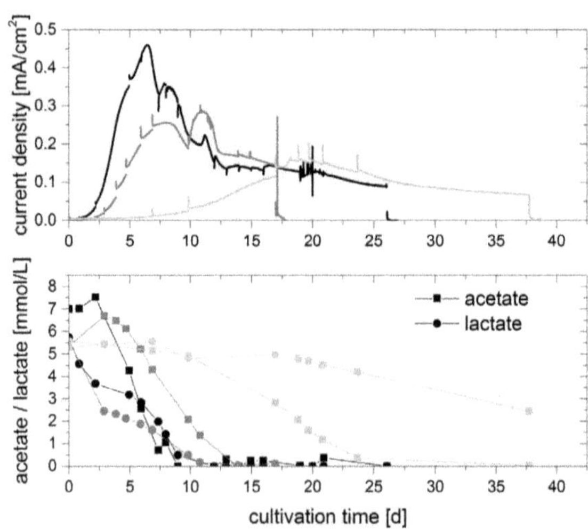

Figure A.5: Chronoamperograms and substrate concentrations of cultivation **cycle 1** of defined mixed cultures of *G. sulfurreducens* and *S. oneidensis* grown at 30 °C on graphite electrodes poised to **0.4** $\mathbf{V_{Ag/AgCl}}$ with 5 mM of acetate and 5 mM of lactate. Differently coloured lines represent individual data from biological triplicates. Spikes in chronoamperogram are due to sampling.

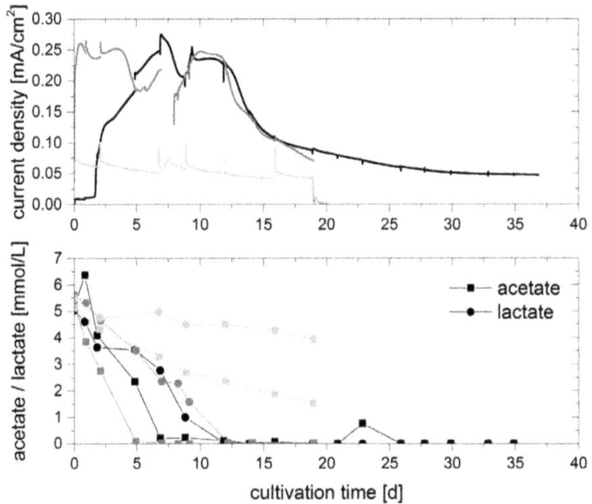

Figure A.6: Chronoamperograms and substrate concentrations of cultivation **cycle 2** of defined mixed cultures of *G. sulfurreducens* and *S. oneidensis* grown at 30 °C on graphite electrodes poised to **0.4 $V_{Ag/AgCl}$** with 5 mM of acetate and 5 mM of lactate. Differently coloured lines represent individual data from biological triplicates. Spikes in chronoamperogram are due to sampling.

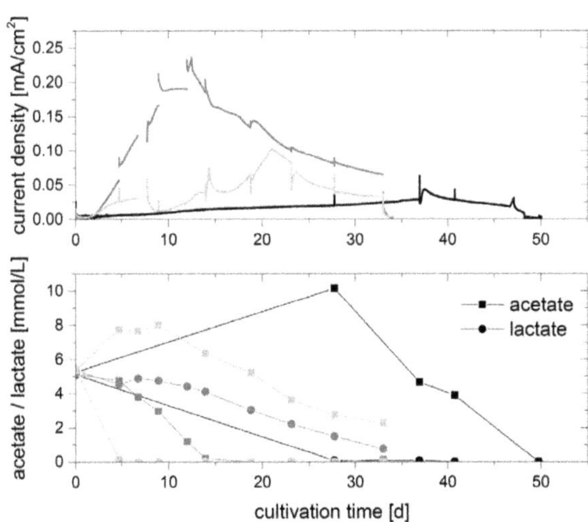

Figure A.7: Chronoamperograms and substrate concentrations of cultivation **cycle 1** of defined mixed cultures of *G. sulfurreducens* and *S. oneidensis* grown at 30 °C on graphite electrodes poised to **0.6 $V_{Ag/AgCl}$** with 5 mM of acetate and 5 mM of lactate. Differently coloured lines represent individual data from biological triplicates. Spikes in chronoamperogram are due to sampling.

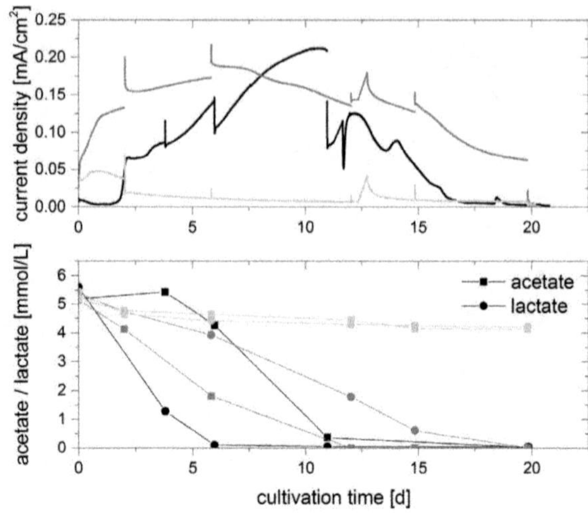

Figure A.8: Chronoamperograms and substrate concentrations of cultivation **cycle 2** of defined mixed cultures of *G. sulfurreducens* and *S. oneidensis* grown at 30 °C on graphite electrodes poised to **0.6 $V_{Ag/AgCl}$** with 5 mM of acetate and 5 mM of lactate. Differently coloured lines represent individual data from biological triplicates. Spikes in chronoamperogram are due to sampling.

Succinate concentration during cultivation cycle 1 of *S. oneidensis* pure culture and of defined mixed cultures at varying anode potentials:

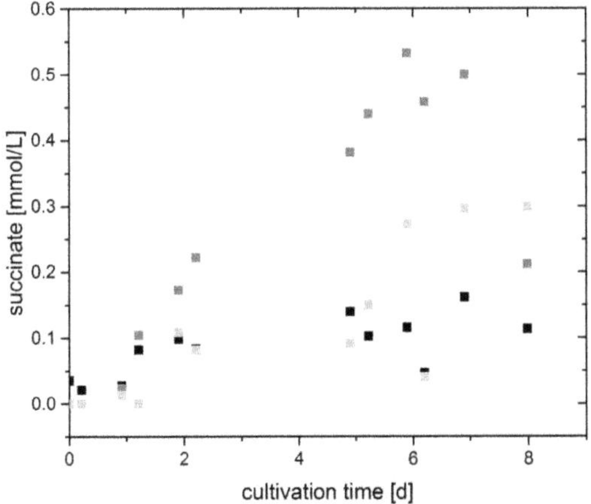

Figure A.9: Succinate concentration of cultivation cycle 1 of ***S. oneidensis* pure culture** grown at 30 °C on graphite electrodes poised to 0.2 $V_{Ag/AgCl}$ with 10 mM of lactate. Different colours represent individual data from biological triplicates.

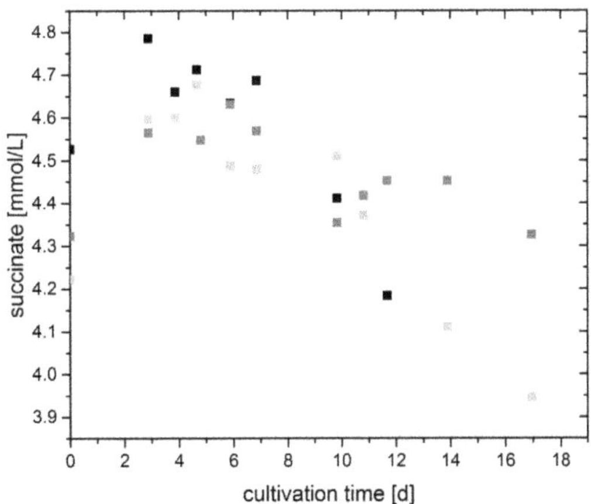

Figure A.10: Succinate concentration of cultivation cycle 1 of defined mixed cultures of *G. sulfurreducens* and *S. oneidensis* grown at 30 °C on graphite electrodes poised to **-0.2 $V_{Ag/AgCl}$** with 10 mM of lactate. Different colours represent individual data from biological triplicates.

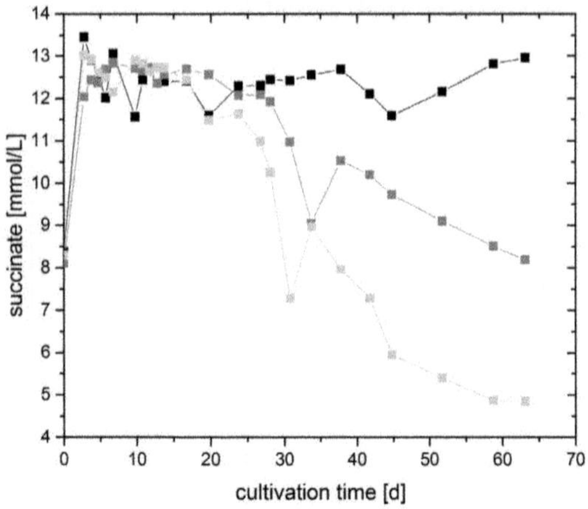

Figure A.11: Succinate concentration of cultivation cycle 1 of defined mixed cultures of *G. sulfurreducens* and *S. oneidensis* grown at 30 °C on graphite electrodes poised to **0.0 $V_{Ag/AgCl}$** with 10 mM of lactate. Different colours represent individual data from biological triplicates.

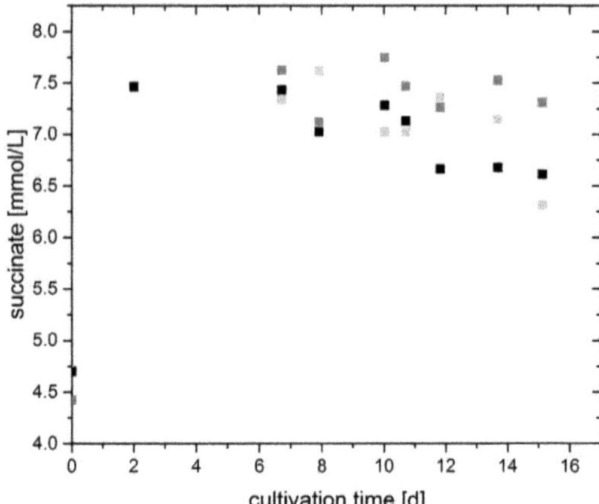

Figure A.12: Succinate concentration of cultivation cycle 1 of defined mixed cultures of *G. sulfurreducens* and *S. oneidensis* grown at 30 °C on graphite electrodes poised to **0.2 $V_{Ag/AgCl}$** with 10 mM of lactate. Different colours represent individual data from biological triplicates.

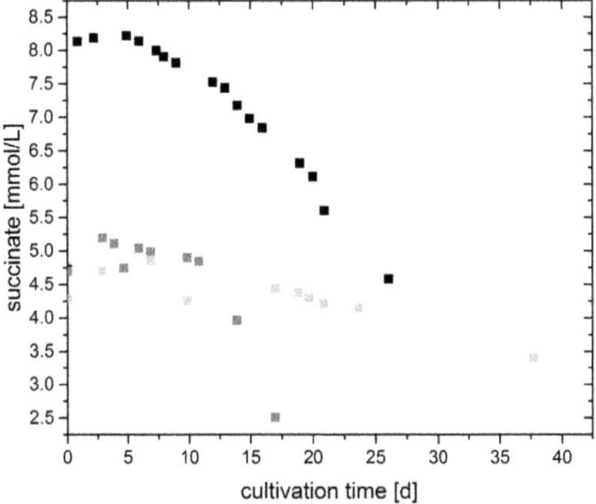

Figure A.13: Succinate concentration of cultivation cycle 1 of defined mixed cultures of *G. sulfurreducens* and *S. oneidensis* grown at 30 °C on graphite electrodes poised to **0.4** $\mathbf{V}_{Ag/AgCl}$ with 10 mM of lactate. Different colours represent individual data from biological triplicates.

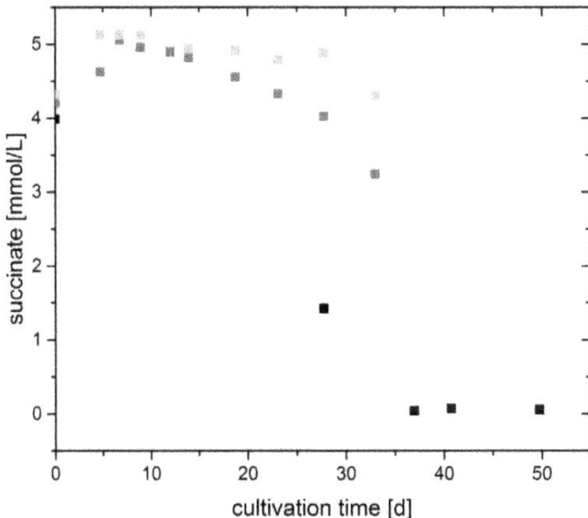

Figure A.13: Succinate concentration of cultivation cycle 1 of defined mixed cultures of *G. sulfurreducens* and *S. oneidensis* grown at 30 °C on graphite electrodes poised to **0.6** $\mathbf{V}_{Ag/AgCl}$ with 10 mM of lactate. Different colours represent individual data from biological triplicates.

Formal redox potentials found in biofilms of defined mixed culture at varying anode potentials and in *G. sulfurreducens* pure culture by cyclic voltammetry with corresponding peak assignment:

Table A.1: List of formal redox potentials and corresponding Peak assignments found by CV in defined mixed culture at varying anode potentials (vs. Ag/AgCl) and in *G. sulfurreducens* pure culture (*G. sulf*). Grey area indicates potentials not covered in *G. sulfurreducens* pure culture CV.

Assignment	-0.2 V	0.0 V	0.2 V	0.4 V	0.6 V	*G. sulf*
A	-0.4253	-0.4164	-0.4213		-0.4099	
B	-0.3518	-0.3558	-0.3657	-0.3650	-0.3401	-0.3635
C	-0.3448	-0.3223	-0.3311	-0.2976	-0.2443	-0.3266
D	-0.2757	-0.2782	-0.2656	-0.2411	-0.2193	-0.2637
E	-0.2185	-0.2262	-0.2147	-0.1661	-0.1835	
F		-0.1648		-0.1784		-0.1780
G	-0.1291	-0.1292	-0.1317		-0.1212	-01194
H	-0.0447	-0.0245	-0.0650	-0.0400	-0.0575	
I	0.0508	0.0645	0.0382	-0.0020	0.0253	0.0241
J	0.1427		0.1279	0.1414	0.0986	
K	0.2040	0.1691	0.1784		0.2063	
L			0.2651			
M	0.3419	0.3001	0.3121	0.3520	0.3242	
N			0.3760		0.3703	
O					0.5066	

Band 1 **Sunder, Matthias**: Oxidation grundwasserrelevanter Spurenverunreinigungen mit Ozon und Wasserstoffperoxid im Rohrreaktor. 1996. FIT-Verlag · Paderborn, ISBN 3-932252-00-4

Band 2 **Pack, Hubertus**: Schwermetalle in Abwasserströmen: Biosorption und Auswirkung auf eine schadstoffabbauende Bakterienkultur. 1996. FIT-Verlag · Paderborn, ISBN 3-932252-01-2

Band 3 **Brüggenthies, Antje**: Biologische Reinigung EDTA-haltiger Abwässer. 1996. FIT-Verlag · Paderborn, ISBN 3-932252-02-0

Band 4 **Liebelt, Uwe**: Anaerobe Teilstrombehandlung von Restflotten der Reaktivfärberei. 1997. FIT-Verlag · Paderborn, ISBN 3-932252-03-9

Band 5 **Mann, Volker G.**: Optimierung und Scale up eines Suspensionsreaktorverfahrens zur biologischen Reinigung feinkörniger, kontaminierter Böden. 1997. FIT-Verlag · Paderborn, ISBN 3-932252-04-7

Band 6 **Boll Marco**: Einsatz von Fuzzy-Control zur Regelung verfahrenstechnischer Prozesse. 1997. FIT-Verlag · Paderborn, ISBN 3-932252-06-3

Band 7 **Büscher, Klaus**: Bestimmung von mechanischen Beanspruchungen in Zweiphasenreaktoren. 1997. FIT-Verlag · Paderborn, ISBN 3-932252-07-1

Band 8 **Burghardt, Rudolf**: Alkalische Hydrolyse – Charakterisierung und Anwendung einer Aufschlußmethode für industrielle Belebtschlämme. 1998. FIT-Verlag · Paderborn, ISBN 3-932252-13-6

Band 9 **Hemmi, Martin**: Biologisch-chemische Behandlung von Färbereiabwässern in einem Sequencing Batch Process. 1999. FIT-Verlag · Paderborn, ISBN 3-932252-14-4

Band 10 **Dziallas, Holger**: Lokale Phasengehalte in zwei- und dreiphasig betriebenen Blasensäulenreaktoren. 2000. FIT-Verlag · Paderborn, ISBN 3-932252-15-2

Band 11 **Scheminski, Anke**: Teiloxidation von Faulschlämmen mit Ozon. 2001. FIT-Verlag · Paderborn, ISBN 3-932252-16-0

Band 12 **Mahnke, Eike Ulf**: Fluiddynamisch induzierte Partikelbeanspruchung in pneumatisch gerührten Mehrphasenreaktoren. 2002. FIT-Verlag · Paderborn, ISBN 3-932252-17-9

Band 13 **Michele, Volker**: CDF modeling and measurement of liquid flow structure and phase holdup in two- and three-phase bubble columns. 2002. FIT-Verlag · Paderborn, ISBN 3-932252-18-7

Band 14 **Wäsche, Stefan**: Einfluss der Wachstumsbedingungen auf Stoffübergang und Struktur von Biofilmsystemen. 2003. FIT-Verlag · Paderborn, ISBN 3-932252-19-5

Band 15 **Krull Rainer**: Produktionsintegrierte Behandlung industrieller Abwässer zur Schließung von Stoffkreisläuren. 2003. FIT-Verlag · Paderborn, ISBN 3-932252-20-9

Band 16 **Otto, Peter**: Entwicklung eines chemisch-biologischen Verfahrens zur Reinigung EDTA enthaltender Abwässer. 2003. FIT-Verlag · Paderborn, ISBN 3-932252-21-7

Band 17 **Horn, Harald**: Modellierung von Stoffumsatz und Stofftransport in Biofilmsystemen. 2003. FIT-Verlag · Paderborn, ISBN 3-932252-22-5

Band 18 **Mora Naranjo, Nelson**: Analyse und Modellierung anaerober Abbauprozesse in Deponien. 2004. FIT-Verlag · Paderborn, ISBN 3-932252-23-3

Band 19 **Döpkens, Eckart**: Abwasserbehandlung und Prozesswasserrecycling in der Textilindustrie. 2004. FIT-Verlag · Paderborn, ISBN 3-932252-24-1

Band 20 **Haarstrick, Andreas**: Modellierung millieugesteuerter biologischer Abbauprozesse in heterogenen problembelasteten Systemen. 2005. FIT-Verlag · Paderborn, ISBN 3-932252-27-6

Band 21 **Baaß, Anne-Christina**: Mikrobieller Abbau der Polyaminopolycarbonsäuren Propylendiamintetraacetat (PDTA) und Diethylentriaminpentaacetat (DTPA). 2004. FIT-Verlag · Paderborn, ISBN 3-932252-26-8

Band 22 **Staudt, Christian**: Entwicklung der Struktur von Biofilmen. 2006. FIT-Verlag · Paderborn, ISBN 3-932252-28-4

Band 23 **Pilz, Roman Daniel**: Partikelbeanspruchung in mehrphasig betriebenen Airlift-Reaktoren. 2006. FIT-Verlag · Paderborn, ISBN 3-932252-29-2

Band 24 **Schallenberg, Jörg**: Modellierung von zwei- und dreiphasigen Strömungen in Blasensäulenreaktoren. 2006. FIT-Verlag · Paderborn, ISBN 3-932252-30-6

Band 25 **Enß, Jan Hendrik**: Einfluss der Viskosität auf Blasensäulenströmungen. 2006. FIT-Verlag · Paderborn, ISBN 3-932252-31-4

Band 26 **Kelly, Sven**: Fluiddynamischer Einfluss auf die Morphogenese von Biopellets filamentöser Pilze. 2006. FIT-Verlag · Paderborn, ISBN 3-932252-32-2

Band 27 **Grimm, Luis Hermann**: Sporenaggregationsmodell für die submerse Kultivierung koagulativer Myzelbildner. 2006. FIT-Verlag · Paderborn, ISBN 3-932252-33-0

Band 28 **León Ohl, Andrés**: Wechselwirkungen von Stofftransport und Wachstum in Biofilsystemen. 2007. FIT-Verlag · Paderborn, ISBN 3-932252-34-9

Band 29 **Emmler, Markus**: Freisetzung von Glucoamylase in Kultivierungen mit *Aspergillus niger*. 2007. FIT-Verlag · Paderborn, ISBN 3-932252-35-7

Band 30 **Leonhäuser, Johannes**: Biotechnologische Verfahren zur Reinigung von quecksilberhaltigem Abwasser. 2007. FIT-Verlag · Paderborn, ISBN 3-932252-36-5

Band 31 **Jungebloud, Anke**: Untersuchung der Genexpression in *Aspergillus niger* mittels Echtzeit-PCR. 1996. FIT-Verlag · Paderborn, ISBN 978-3-932252-37-2

Band 32 **Hille, Andrea**: Stofftransport und Stoffumsatz in filamentösen Pilzpellets. 2008. FIT-Verlag · Paderborn, ISBN 978-3-932252-38-9

Band 33 **Fürch, Tobias**: Metabolic characterization of recombinant protein production in *Bacillus megaterium*. 2008. FIT-Verlag · Paderborn, ISBN 978-3-932252-39-6

Band 34 **Grote, Andreas Georg**: Datenbanksysteme und bioinformatische Werkzeuge zur Optimierung biotechnologischer Prozesse mit Pilzen. 2008. FIT-Verlag · Paderborn, ISBN 978-3-932252-40-120

Band 35 **Möhle, Roland Bernhard**: An Analytic-Synthetic Approach Combining Mathematical Modeling and Experiments – Towards an Understanding of Biofilm Systems. 2008. FIT-Verlag · Paderborn, ISBN 978-3-932252-41-9

Band 36 **Reichel, Thomas**: Modelle für die Beschreibung das Emissionsverhaltens von Siedlungsabfällen. 2008. FIT-Verlag · Paderborn, ISBN 978-3-932252-42-6

Band 37 **Schultheiss, Ellen**: Charakterisierung des Exopolysaccharids PS-EDIV von *Sphingomonas pituitosa*. 2008. FIT-Verlag · Paderborn, ISBN 978-3-932252-43-3

Band 38 **Dreger, Michael Andreas**: Produktion und Aufarbeitung des Exopolysaccharids PS-EDIV aus *Sphingomonas pituitosa*. 1996. FIT-Verlag · Paderborn, ISBN 978-3-932252-44-0

Band 39 **Wiebels, Cornelia**: A Novel Bubble Size Measuring Technique for High Bubble Density Flows. 2009. FIT-Verlag · Paderborn, ISBN 978-3-932252-45-7

Band 40 **Bohle, Kathrin**: Morphologie- und produktionsrelevante Gen- und Proteinexpression in submersen Kultivierungen von *Aspergillus niger*. 2009. FIT-Verlag · Paderborn, ISBN 978-3-932252-46-2

Band 41 **Fallet, Claas**: Reaktionstechnische Untersuchungen der mikrobiellen Stressantwort und ihrer biotechnologischen Anwendungen. 2009. FIT-Verlag · Paderborn, ISBN 978-3-932252-47-1

Band 42 **Vetter, Andreas**: Sequential Co-simulation as Method to Couple CFD and Biological Growth in a Yeast. 2009. FIT-Verlag · Paderborn, ISBN 978-3-932252-48-8

Band 43 **Jung, Thomas**: Einsatz chemischer Oxidationsverfahren zur Behandlung industrieller Abwässer. 2010. FIT-Verlag· Paderborn, ISBN 978-3-932252-49-5

Band 45 **Herrmann, Tim**: Transport von Proteinen in Partikeln der Hydrophoben Interaktions Chromatographie. 2010. FIT-Verlag · Paderborn, ISBN 978-3-932252-51-8

Band 46 **Becker, Judith**: Systems Metabolic Engineering of *Corynebacterium glutamicum* towards improved Lysine Prodction. 2010. Cuvillier-Verlag · Göttingen, ISBN 978-3-86955-426-6

Band 47 **Melzer, Guido**: Metabolic Network Analysis of the Cell Factory *Aspergillus niger*. 2010. Cuvillier-Verlag · Göttingen, ISBN 978-3-86955-456-3

Band 48 **Bolten J., Christoph**: Bio-based Production of L-Methionine in *Corynebacterium glutamicum*. 2010. Cuvillier-Verlag · Göttingen, ISBN 978-3-86955-486-0

Band 49 **Lüders, Svenja**: Prozess- und Proteomanalyse gestresster Mikroorganismen. 2010. Cuvillier-Verlag · Göttingen, ISBN 978-3-86955-435-8

Band 50 **Wittmann, Christoph**: Entwicklung und Einsatz neuer Tools zur metabolischen Netzwerkanalyse des industriellen Aminosäure-Produzenten *Corynebacterium glutamicum*. 2010. Cuvillier-Verlag · Göttingen, ISBN 978-3-86955-445-7

Band 51 **Edlich, Astrid**: Entwicklung eines Mikroreaktors als Screening-Instrument für biologische Prozesse. 2010. Cuvillier-Verlag · Göttingen, ISBN 978-3-86955-470-9

Band 52 **Hage, Kerstin**: Bioprozessoptimierung und Metabolomanalyse zur Proteinproduktion in *Bacillus licheniformis*. 2010. Cuvillier-Verlag · Göttingen, ISBN 978-3-86955-578-2

Band 53 **Kiep, Katina Andrea**: Einfluss von Kultivierungsparametern auf die Morphologie und Produktbildung von *Aspergillus niger*. 2010. Cuvillier-Verlag · Göttingen, ISBN 978-3-86955-632-1

Band 54 **Fischer, Nicole**: Experimental investigations on the influence of physico-chemical parameters on anaerobic degradation in MBT residual waste. 2011. Cuvillier-Verlag · Göttingen, ISBN 978-3-86955-679-6

Band 55 **Schädel, Friederike**: Stressantwort von Mikroorganismen. 2011. Cuvillier-Verlag · Göttingen, ISBN 978-3-86955-746-5

Band 56 **Wichter, Johannes**: Untersuchung der L-Cystein-Biosynthese in *Escherichia coli* mit Techniken der Metabolom- und ^{13}C-Stofffflussanalyse. 2011. Cuvillier-Verlag · Göttingen, ISBN 978-3-86955-750-2

Band 57 **Knappik, Irena Isabell**: Charakterisierung der biologischen und chemischen Reaktionsprozesse in Siedlungsabfällen. 2011. Cuvillier-Verlag · Göttingen, ISBN 978-3-86955-760-1

Band 58 **Driouch, Habib**: Systems biotechnology of recombinant protein production in *Aspergillus niger*. 2011. Cuvillier-Verlag Göttingen, ISBN 978-3-86955-808-0

Band 59 **Gehder, Matthias:** Development and Validation of Indicators for the Production and Quality of Seed Cultures. 2011. Cuvillier-Verlag Göttingen, ISBN 978-3-86955-847-9

Band 60 **Sommer, Becky:** Methodenentwicklung zur Charakterisierung sporenbildender Pilz-Seedingkulturen. 2011. Cuvillier-Verlag Göttingen, ISBN 978-3-86955-851-6

Band 61 **Dohnt, Katrin:** Charakterisierung von *Pseudomonas aeruginosa*-Biofilmen in einem *in vitro*-Harnwegskathetersystem. 2011. Cuvillier-Verlag Göttingen, ISBN 978-3-86955-852-3

Band 62 **Greis, Tillman:** Meddling the risk of chlorinated hydrocarbons in urban groundwater. 2011. Cuvillier-Verlag Göttingen, ISBN 978-3-86955-970-4

Band 63 **David, Florian:** Holistic bioprocess engineering of antibody fragment secreting *Bacillus megaterium*. 2012. Cuvillier-Verlag Göttingen, ISBN 978-3-95404-115-2

Band 64 **Palme, Wiebke:** Taxonomische Einordnung des Polyaminopolycarbonsäure-abbauenden Stammes BNC1 und Untersuchungen zum Abbau von 1,3 Propylendiamintetraacetat. 2012. Cuvillier-Verlag Göttingen, ISBN 978-3-95404-158-9

Band 65 **Lin, Pey-Jin:** Effect of fluid dynamics on pellet morphology and product formation of *Aspergillus niger*. 2012. Cuvillier-Verlag Göttingen, ISBN 978-3-95404-181-7

Band 66 **Kind, Stefanie:** Synthetic Metabolic Engineering of *Corynebacterium glutamicum* for Bio-based Production of 1,5-Diaminopentane. 2012. Cuvillier-Verlag Göttingen, ISBN 978-3-95404-264-7

Band 67 **Wilk, Franziska:** Charakterisierung der Stoffströme vorbehandelter Siedlungsabfälle in Deponiebioreaktoren. 2012. Cuvillier-Verlag Göttingen, ISBN 978-3-95404-281-4

Band 68 **Korneli, Claudia:** Target-oriented Bioprocess Optimization for Recombinant Protein Production in *Bacillus megaterium*. 2012. Cuvillier-Verlag Göttingen, ISBN 978-3-95404-289-0

Band 69 **Eslahpazir, Manely:** Numerical Characterization of Mechanical Stress and Flow Patterns in Stirred Tank Bioreactors. 2013. Cuvillier-Verlag Göttingen, ISBN 978-3-95404-449-8

Band 70 **Wucherpfennig, T.:** Cellular Morphology – A novel Process Parameter for the Cultivation of Eukaryotic Cells. 2013. Cuvillier-Verlag Göttingen, ISBN 978-3-95404-456-6

Band 71 **Buschke, Nele:** Bio-Nylon Monomers from Renewables using *Corynebacterium glutamicum*. 2013. Cuvillier-Verlag Göttingen, ISBN 978-3-95404-457-3

Band 72 **Bergmann, Sven:** Ectoine production by halotolerant microorganisms. 2013. Cuvillier-Verlag Göttingen, ISBN 978-3-95404-556-3

Band 73 **Hellriegel, Jan:** Engineering a Biofilm – Imitating Physico-Chemical Properties to improve Mechanical Characterization. 2014. Cuvillier-Verlag Göttingen, ISBN 978-3-95404-753-6

Band 74 **Berger, Antje:** Metabolische Netzwerkanalyse uropathogener *Pseudomonas aeruginosa*-Isolate. 2014. Cuvillier-Verlag Göttingen, ISBN 978-3-95404-762-8

Band 75 **Peterat, Gena:** Prozesstechnik und rekationskinetische Analysen in einem mehrphasigen Mikrobioreaktorsystem. 2014. Cuvillier-Verlag Göttingen, ISBN 978-3-95404-887-8

Band 76 **Godard, Thibault:** Systems biology of stress in *Bacillus megaterium* and its potential applications. 2016. Cuvillier-Verlag Göttingen, ISBN 978-3-7369-9336-5

Band 77 **Hönnscheidt, Christoph:** Entwicklung kolloiddisperser Wirkstoffformulierungen auf Basis von Biopolymeren. 2016. Cuvillier-Verlag Göttingen, ISBN 978-3-7369-9260-3

Band 78 **Walisko, Jana:** Morphologiebeeinflussung von *Lechevalieria aerocolonigenes* und heterologe Produktion von Rebeccamycin. 2017. Cuvillier-Verlag Göttingen, ISBN 978-3-7369-9502-4

Band 79 **Gädke, Johannes:** *In situ*-downstream processing of recombinant histidine-tagged proteins from cultivations of *Bacillus megaterium*. 2017. Cuvillier-Verlag Göttingen, ISBN 978-3-7369-9551-2

Band 80 **Lakowitz, Antonia:** Skalenübergreifende Produktion und Sekretion rekom--binanter Proteine mit Stämmen der Gattung *Bacillus*. 2017. Cuvillier-Verlag Göttingen, ISBN 978-3-7369-9576-5

Band 81 **Lladó Maldonado, Susanna Maria:** Bioengineering at the micro-scale: Design, characterization and validation of microbioreactors. 2019. Cuvillier-Verlag Göttingen, ISBN 978-3-7369-7025-0

www.ingramcontent.com/pod-product-compliance
Ingram Content Group UK Ltd.
Pitfield, Milton Keynes, MK11 3LW, UK
UKHW022000190726
13853UKWH00004B/1634

9 783736 972117